THE EARTH MOVING AROUND
WHAT PROMOTES THAT?

NELSON DAVID

KENSYESS3
TYLER, TX

The Earth Moving Around,
What Promotes that?
Nelson David

Published by Kensyess3

Library of Congress Control Number: 2021909108

ISBN 978-1-7333749-5-8 (paperback)
ISBN 978-1-7333749-6-5 (ebook)

To my wife Nelly, my daughter Yessika, to my mother Petra's memory, and to Luis, Mary, Xio, Eva, Eli, Roger, and Merly.

Contents

Preface

> You can vanish but still are here, because my memory and affection trapped you inside of me.
>
> Merly C. Ojeda

Are there in your mind things or thoughts that have disturbed you for years?

Maybe and probably, one of them is the law of universal gravitation. I was captivated with this law since the first time I heard about it. Simply put, it is when the Earth is suspended, hung, or attracted by one force of gravitation without falling, and is rotating, and orbiting around the Sun. Such a strange, yet beautiful phenomenon this law is. It was discovered by one of the most brilliant scientists of all the time, the great Isaac Newton. We must take for granted this law stated by such a genius, and we do but why is it so disturbing for many of us, the common person? This law also upset another great scientist, Albert Einstein, who said "there is no such thing as gravitational force (Serway and Jewett 1141)."

The Earth is attracted to the Sun but they do not collide. It is amazing we are not flying through other galaxies despite the weak gravitational force between Mother Earth and the Sun. This synergistic coupling begs the following questions. Why does our planet rotate and orbit and what promotes these movements? Does only one force promote these movements? Why is Earth's orbit on an elliptical and almost circular trajectory?

If you want answers to these questions, wonder why the Earth is moving around, and are open to a new hypothesis or a different point of view about the law of universal gravitation, then this book is for you.

This law is stated as "Every particle in the Universe attracts every other particle with *a force* that is directly proportional to the product of their masses and inversely proportional to the square of the distance between them" (Serway and Jewett 363). On this text, based on the results of homemade experiments, I am going to demonstrate that instead of this unique force there are three forces acting over the Earth, one in the direction of the Earth's orbit, another acting away from the Sun, and counter balanced by the third one aimed toward the Sun. It is also proved that such orbital force is generated by an imbalance in the Earth and by its angular momentum. These two factors also make our planet spin.

We are taught that the Earth is kept "linked" to the Sun by the gravitational force. However, this established way of thinking changed for me during a fortuitous playing with a coin on the kitchen counter-top. That episode triggered and unchained most of the experiments that you will read in this book, all of which are presented on a logical sequence.

Acknowledgments

My thanks to my daughter Yessika Guerra for her immeasurable support to shaping the manuscript from its first bundle of scrawled notes, and my wife for her patience to help me to make the book unravel.

I also thank the following people for their thoughtful comments: Edgar Marin, Nelson Fumo, Lyneisa Ray-Davison, Felix Cestari, Joy Alexander, Jack Jenkins, and Dr. Ijeoma Unegbu for accuracy in proofreading.

THE EARTH MOVING AROUND

Introduction

Imagination is more important than knowledge. Knowledge is limited, imagination encircles the world.
Albert Einstein

The book consists of two macro-sections: the first one, the most important, is formed by four parts. The first has a group of definitions of physical concepts that are needed to describe and interpret the experiments. These concepts that we, aware or not, have used hands-on through the years. The second part summarizes the experiments performed with rotating cylindrical objects (ring, disk, wheel), and it also contains a summary stemming from a documentary about a spinning cylinder. The third part is a collection of experiments done with air-suspended round bodies (balls), and the last, that is the core of the book, explains the association between the experiments and the orbital and rotational movements of our planet Earth. The second macro-section is just a complement to the first, and is composed of three parts that details the experiments of the first section, and others associated to them and the Earth. Also included is a revision of mathematical topics needed to determine physical properties of the utilized objects. In these two last parts is where we really go a little deeper on the mathematics.

Most of the sections of this book are mainly related to simple and home-made experiments (tests to demonstrate or confirm mostly well-known facts) that can be made using common and handy things. The do-it-yourself experiments described or presented in further sections are linked and related. If you try some of them, remember that "The science behind the activities doesn't really matter (okay, the science is important), what truly matters is the experience" (Spangler 6). As it was stated by John Hunter, who is considered the father of modern medicine, "Don't think; try the experiment" (Usborne, Sykes and Bee, Episode 2: A Roomful of Brilliant Minds, Chapter 3: John Hunter). Most of the experiments performed are popular and well-known, but are presented with subtle changes, and/or deepen on uncommon aspects, or whose results are viewed from a different point of view. These experiments are centered in physics whose natural and main mathematical language has been reduced to sketches, graphics, and simple examples for improved reader understanding of critical concepts and ideas introduced in the text. They can be reproduced to validate the results and to show such key ideas in action.

The moment of inertia, and other properties and mathematical formulas used in the book can be found in popular literature or on the internet. However, the development of some of them are exposed because in some cases it is not easy to find them (like the moment of inertia of a common key ring), or in other cases, because a known formula is given in terms which are different from those required on a particular application.

Part A A spoonful of definitions

If you thought that science was certain – Well, that is just an error on your part.
Richard Feynman

We normally have a feeling or a sense of most of the definitions detailed in the next paragraphs, however, we usually do not know them by their scientific name, or we do not know how they are presented or drawn on sketches. Please become familiar with these definitions because they not only are frequently mentioned in future sections, but they also allow us to describe and interpret the heart of the book and the homemade experiments and their results.

Force, vector, and torque

To generate a pushing or pulling effect we must exert an action called *force*. How pushing and pulling are different effects, in physics and math, a direction is given to them to differentiate one from the other. Anything that needs at least two characteristics to clearly identify its effect, is or could be defined as a *vector* representing these traits. For instance, the distance (magnitude) between two points is a scalar quantity, but if we use one of its points as reference and give a direction (left, right, up, etc.) to the other, we transform the distance into a position vector. So, a force is a vector that is usually represented with a (F) letter and a straight arrow, whose point end or head aims in the direction that we push or pull.

When we travel at the same speed and on the same straight road first to the north and then to the south of the town, we are moving (not simultaneously) with two different velocities. Velocity, as well as force, is a vector. It has magnitude (speed) and direction (north or south). To visualize a vector, imagine it as a bow arrow which length is its magnitude and the pointing head shows its direction of action.

We can add or subtract similar vectors (force to force, torque to torque, etc.) acting simultaneously on a spot or (*P*) point. To graphically do that, imagine arrows (for instance two) of different size and aiming in different directions. For example, let's call **α** one such vectors and **β** the other one, see Figure A1a. Keeping the original direction and magnitude of the vectors, to add them place the tail of one of them on the head of the other. The addition is as follows: translate the tail of one of them (**α**), let's call it the first, to the point P when one of them is not already in such spot. Then, or if the head or tail of the first is already in such point, just translate to the head of the first one, the tail of the other (**β** vector) arrow to get the addition (**α** + **β**) of vector **β** to vector **α**. The magnitude and direction of the addition of these vectors goes from the tail of the first one to the head of the second one. See "red" vector or vector ***A*** in Figure A1b.

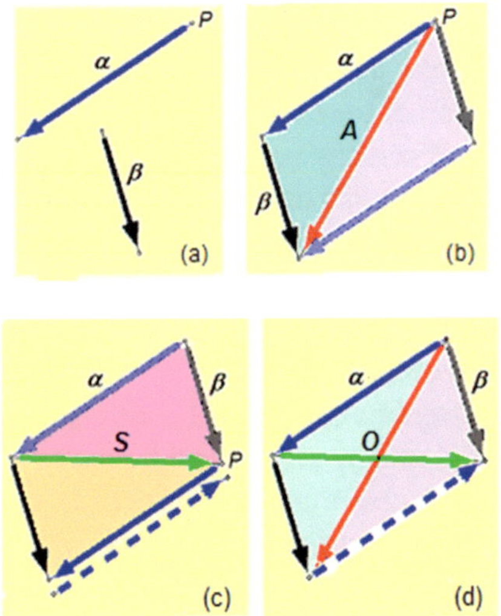

Figure A1. Addition and subtraction of vectors

Notice that when the two vectors are not parallel, nor collinear or acting in a same line, they and the resulting vector form a triangle, as the blue one shown in Figure A1b. On the other hand, when the vector **α** is the one added to vector **β**, the (**β** + **α**) result vector is also the same (***A***). However in this case, the sum forms the purple triangle. These "two" additions or two triangles form a parallelogram that is a quadrilateral whose pair of opposite sides are parallel and have the same length or magnitude. Each of the opposite sides with same magnitude is the same vector once with its tail in the P point and the second time in the head of the other vector.

Similarly, to subtract two vectors, keeping their direction and magnitude, place them head to head or tail to tail to get the magnitude of the result vector. However, the result vectors of **α** – **β** and **β** – **α** subtractions have opposite directions. To know the subtraction direction, on the parallelogram form by the head-to-head or the tail-to-tail subtractions, see Figure A1c, make "negative" the vector we want to subtract. This negativism makes such vector aim on the opposite direction or swaps its head and tail (see dashed arrow in Figure A1c). After that, the subtractions become the addition of the original vector to the now "negative" [1] one. For example, if **α** is the swapped (negative) vector, in either the tail-to-tail (pink) triangle or head-to-head (orange) triangle, the subtraction becomes the addition of **-α** + **β** or **β** – **α**, whose direction of the result ***S*** vector is the green one (left to right) shows by Figure A1c. If **β** is the negative in either triangle, the subtraction direction has to be opposite.

The parallelograms of Figure A1c and Figure A1d actually are a unique one because they are formed by the same vectors. So, if the ***A*** and ***S*** result vectors are included in it, they become the diagonals of such parallelogram, see Figure A1d. The parallelograms have an interesting property, their diagonals (or

---

[1] The "negative" is really other vector aiming in the opposite direction but having same magnitude as its opposite one.

the results ***A*** and ***S*** vectors) bisect each other[2]. On Figure A1d, it means that the point *O* is the midpoint of both the subtraction result vector ***S*** and the addition result vector ***A***.

Figure A2 shows an actual case of addition of vectors. The vertical tension of the rope supporting the hanging pail is transferred through a pulley to the inclined section of the rope attached to the wall. The tension (*T*) in this section has a component (vector) aiming down, and another horizontal, aiming to the left wall, that makes the pulley be offset of its hanging point (*A*). To see additional tensions and forces acting in this pulley see Figure C5. For additional examples of vector additions see Figure B6c for addition and subtractions of collinear velocities, and Figure D8 for addition of collinear and perpendicular vectors.

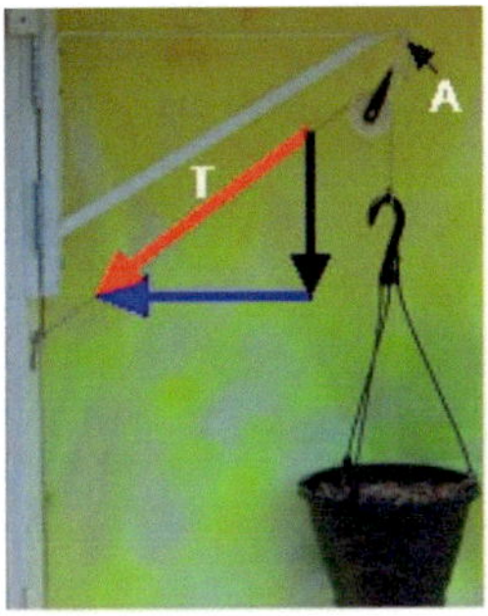

Figure A2. Vectors and a hanging pail

Figure A3a shows a typical heavy, non-automatic door common on a shopping center. To open the door one has to conquer a closing resistance exerted by the device that always moves the door toward the closed position[3]. As the resistance of the device is almost constant, we have to push or pull harder when we do any of these actions on a (n) spot near to the hinges than when we apply our effort (force) at the door handle, which is far

---

[2] The proof of this principle can be found in a good book of geometry.

[3] A common house door well maintained usually offers a small resistance to be opened.

from the hinges. A force direction and its magnitude normally are represented by a letter *f* or *F* as it is shown in Figure A3, and also by the letter *T* when the pulling is exerted by a flexible component like a twine, a fishing line, or a rope like the one of the hanging pail of Figure A2.

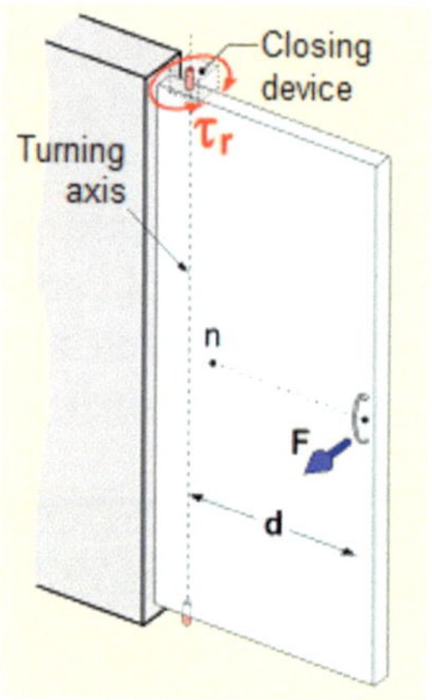

Figure A3. Torque on a swinging door

The effect of a force (F) applied to a certain distance (d) from a point or axis (hinges) is known as *torsion, torque, moment, or moment of a force*. This effect tries to twist, turn, swing, rotate, spin, whirl, or bend the object that usually "feels" or resists the magnitude of this (force times the distance) combination at, or near the turning axis. To differentiate, if we are opening or closing the door, a direction is given to these effects. Thus, a torque is also a vector and is usually represented by the Greek letter tau (τ) drawn over the turning (real or imaginary) axis. See Figure A3. Let's see the door example with numbers: If the resistance (τr) of the closing device is 32 units of torque[4], at a distance of two units of

[4] Typical units of torque are *newton meter* (N.m) when the force is given in *newton* (N) units and the distance in meters (m), and *pound-force foot*

length one needs to apply a force of 16 units (2x16 = 32), and at eight units of distance the required force is just four units of force (8x4 = 32). When we just know the turning direction and the axis affected, this axis and a curved arrow are used to represent the torsion. The curved arrow of a torque is usually drawn with one solid line to differentiate it from the spinning arrow which is drawn with two consecutive lines of different thickness. They could also be differentiated using another approach.

When an applied force is not counteracted by another of equal magnitude and opposite direction on the same line of action, the affected body will move or will be moving in a straight line if the force is strong enough and acting in the direction of the center of mass (see next section). When the force direction is offset from the center of mass, the body will move or will move with a rotational movement about its center of mass. A moving object when disturbed by an external force not strong enough will exert an opposite force of the same magnitude when the disturbance is aligned with its center of mass, and will generate a moment if it is offset from its center of mass.

Energy, mass, and center of masses

Energy is everything that keeps or tries to keep a body or object compact, linked, static, moving and/or alive. During its interchange or transformation, whatever its form, it is searching for minimum consumption, stabilization, and equilibrium. That can be static (Figure A4 and Figure A5b) or dynamic (Figure A5a) in relation to the Earth's surface because we are rotating with Earth and orbiting around the Sun.

Mass is an inherent property of an object or body that allows us to perceive and quantify matter by measuring its volume, weight, and other traits. According to Einstein's theory of relativity, mass should also be considered as energy instead of just a

---

(Lbsf.ft) when the force is pound-force (Lbsf) and the distance is in foot (ft).

property[5], and mass and velocity are both a form of energy. Mass is a storage or agglomeration of energy that when fragmented at the atomic level is released as heat.

Center of mass (CM) is a point where the mass of a body is concentrated; this point, real or fictitious, is partially or totally surrounded by the body's mass, and may or not touch the body. It is also named *center of gravity* (CG) when the gravity is uniform over the entire object or body. If a static body is hanged or suspended from this point, it is balanced. For instance, the paint can handle, wrapped with a sticky cooking (sealing wrap) paper, shows in Figure A4c. It is in *equilibrium* or *balanced* at an almost perfect horizontal level, when it is hanged from a twine passing through its center of mass. For details on how to calculate a center of mass see *Part E*. On round homogeneous bodies let's call *gravity diameter*[6] the one that goes through the geometry center and the center of gravity.

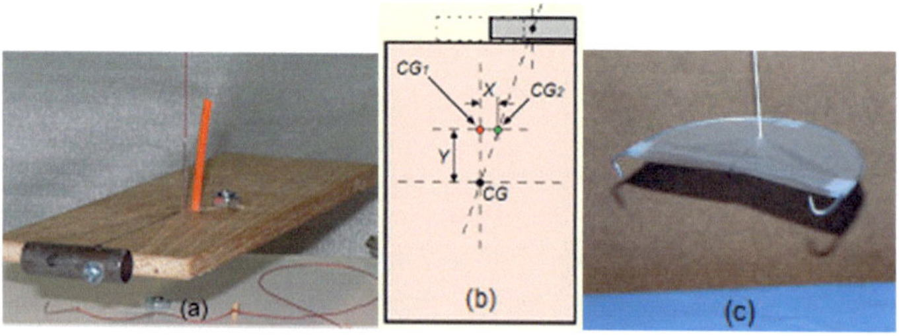

Figure A4. Bodies hanged through center of gravity

The center of mass of combined bodies is located over the line interconnecting the location of the centers of gravity of the joined bodies. It is close to the center of mass of the one having

[5] A property is an attribute, quality, characteristic, etc., that has intensity (magnitude) or is measurable.

[6] A diameter is a straight line passing from side to side through the geometric center of a round body. One that divides a circle into two halves or semicircles.

bigger mass before the combination. Let's use as an example, the object composed of a timber board, and a metal bar that can be switched either to the board center longitudinal line (Figure A4a) or to one board corner (Figure A4b). When the midpoint of the bar coincides with the board center line, the board-bar center of gravity, point CG_1 in Figure A4b, remains on the board's center longitudinal line, but moves away a vertical distance Y from the board center of gravity (*CG*) toward the attached bar. When the bar is located on the board corner, the board-bar center of gravity, point CG_2, is moved a horizontal distance X and a vertical distance Y toward the bar. When each of these two arrangements of the board-bar object is suspended from a thread passing through its center of gravity, it keeps itself on horizontal level position even when rotating at low revolutions per minute. When the object was initially suspended from a vertical, or almost vertical thread, the object spun from what appeared to be torsion stress accumulated on the thread or due to the spinning rotation of the Earth.

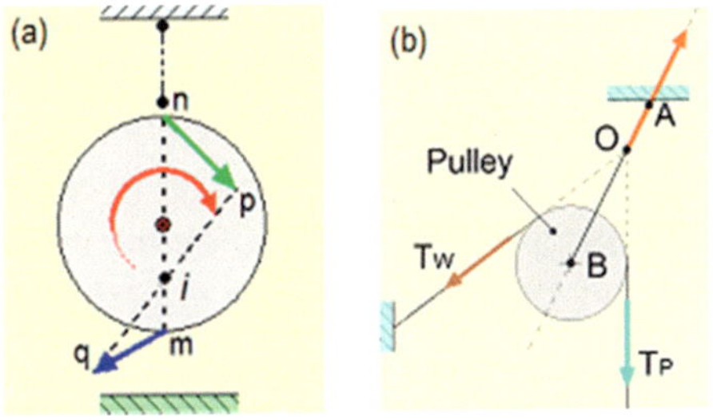

Figure A5. Center of: (a) mass instantaneous (b), equilibrium

When the object is moving, it can also move its center of mass to another location. For instance, the hanged spinning wheel illustrated in Figure A5a has an *instantaneous (i) center of mass*[7] determined by the two vectors (energetic actions) acting in its vertical diameter. This (*i*) point is obtained by intersecting the

[7] Instantaneous center of mass is a concept well-known among engineers and scientists.

imaginary (*n-m*) line passing by the tail of the vectors with the imaginary (*p-q*) line passing through their heads. Either of these lines could be extended when needed.

In some cases, a group of interconnected bodies form a system that may have a point where the encountering forces acting on it are concentrated. For example, the pulley attached to a point *A* through a top (AB) rope shown in Figure A5b. In this system another (bottom) rope, that partially wraps the pulley, supports a heavy pail that hangs in one of its ends while its other end is connected to a wall. When the direction of the two (*Tp* and *Tw*) tensions of the bottom rope are prolonged, they meet in one (O) point located in the top rope. This pulley and the other interconnected (pail, ropes) elements, are in static equilibrium because the two tensions on the bottom rope are in balance at the encounter (O) point with the tension on the top rope.

The squeezing effect, and torque direction

To extract water from a soaking wet towel using our hands, we turn both hands in opposite directions. Some people turn the right-hand knuckles towards the front, others towards their body. To get a *squeezing effect*, it does not matter in what direction you turn the hands. What matters is that they must turn in opposite directions. Whatever the hands' turning direction, the effects are always the same, in that the towel contorts and reduces its diameter approaching the hands towards each other. See Figure A6. The squeezing phenomenon is another example of torque, known as *moment of a couple:* an applied force or couple of parallel forces tending to turn or twist an object that resists the turning action. Usually the resisting object or one of its points is fixed and away from the plane where the force is applied.

The squeezing process agrees with *Newton's third law*, namely that in order to have a prolonged or instantaneous equilibrium, an action shall be resisting one (reaction) acting in opposite direction. In other words, there should be a point where these actions cancel each other out. For a turning hand, the other must turn or resist in the opposite direction. For two torques to be in equilibrium, they should act in different directions towards a common point where they neutralize each other. When there is an imbalance of torques, the body should try or be rotating about an axis passing through such point.

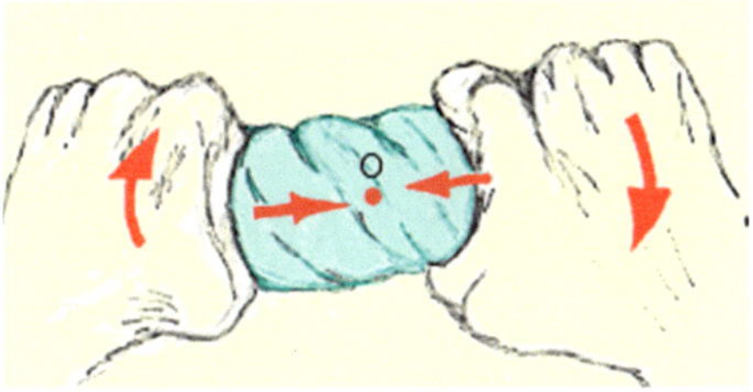

Figure A6. The squeezing effect

Another example of the torque-squeezing couple is found on threaded bolts used to keep two or more pieces or components together. Applying a torque (turning action) on the bolt head and/or on a bolt nut during the tightening process there is a very nil separation between the head and the threaded end (turning reaction) that compresses and tightens the sandwiched pieces. On the pieces-bolt joint is present the action-reaction forces couple. While the pieces are pushing out the bolt, at the same time, its threaded section and molecules are pulling in (tightening) the pieces. Consequently, each particle of a tightened bolt is kept in a continuous pulling and attracting stress.

Regarding the screwing direction, there are two type of screwed bolts, namely, the right-hand thread that are tightened on the clockwise direction, and the left-hand thread tightened on the counterclockwise rotation. Depending on the type used, they can be tightened on the clockwise or on the counterclockwise direction, but in any case, their tightening effect (torque vector direction) is always aimed along and towards the bolted body. However, if we apply the *right-hand rule* to determine the torque vector direction on the tightening of each of type of bolts, we get two opposite directions for the same tightening or squeezing action. This rule gets its name from a right hand partially open and having the thumb extended as it is shown in Figure A7.

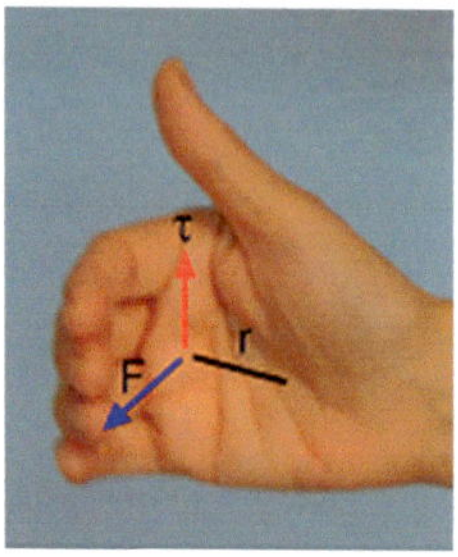

Figure A7. A hand

The hand palm and the fingers are used to represent the direction of the main turning action (force times the distance from a turning axis aligned to the thumb). The palm would be aligned with the (r) distance and the fingers aim in the same direction of the head of the arrow representing the (F) force. The thumb gives the direction of the torque vector (τ). It means that if the fingers (force) would aim in the opposite direction to the one shows by Figure A7 (for instance, going into the book page), you have to turn your wrist accordingly, and the torque (and thumb) direction will be opposed too or aiming down instead of up.

Figure A8. Electric Coil and paper clips

A magnetic field generates a phenomenon similar to the mechanical/physics torque. Let's detail it using a homemade electric coil, an electric wire wound around a thread metal rod, and two paper clips located near the ends of this rod but without touching it as shown in Figure A8. Once the coil is energized by connecting its electric wire ends to a battery, the two paper clips are simultaneously attracted to the central rods by the magnetic field generated. This attraction (squeezing) effect towards the central iron rod will also happen even if the direction of the magnetic field (squeezing direction) is changed by interchanging the wire ends on the battery terminals.

The movement of the clips as if they are compressing the coil rod is like the squeezing effect directed to the center of the body squeezed when torque is applied to an object/axle as described previously. There is an attraction effect that is always directed to the object (center of) mass independent of the squeezing direction in the case of the torque or, current or magnetic field direction of the electric coil.

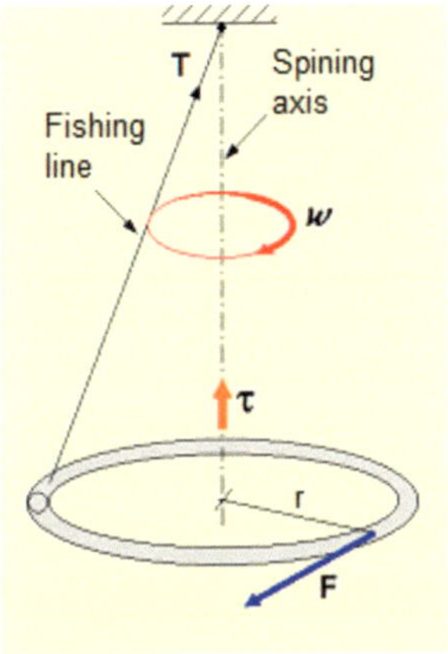

Figure A9. The flat loop lasso spinning

Is it possible that the attracting stress occurring on a tightened bolt may also happen on a "solid" in spinning movement? Yes, for instance on the case of a hanged ring, see Figure A9, the direction of the torque vector (τ) should be visualized to aim towards (upwards in this case) the object, spot, dot, or imaginary line resisting the turning action of the applied torque (product of a radius r and a force F). Notice that the ring moving in this position resembles a cowboy roping trick. The flat loop lasso spins in a horizontal plane, regardless of whether the cowboy moves the lasso clockwise or counterclockwise.

So, this torque direction does not depend on the turning direction (ω) of the moving object receiving the torque action or on the direction of the applied force. In cases like that, the squeezing approach best represents the physical turning-approaching effect, than the accepted but arbitrary right-hand rule for determining the direction of a torque vector. In a hanged spinning ring, using this rule, the torque vector should be aimed upwards or downwards depending on the turning direction of the ring.

Another case of torque direction is found on a bicycle wheel spinning about a horizontal axis while it is hanged just from one end of its axle. For additional details about this topic see Section *The offset-hanging spinning wheel*.

Angular momentum and moment of inertia

When a body is spinning about an axis, the direction of this *rotational velocity* is represented by a curved arrow and its magnitude by a letter than usually is *w*. The red arrow and the letter *w* shown in Figure A9 represent the spinning of the fishing line and the spinning of the hanging ring. In this sketch, it is suggested they are rotating at the same rotational speed and about the same *spinning axis*. The rotational velocity is known by different names depending on the location of the body in regard to the rotating axis.

Figure A10 shows some of these movements and the name of the associated axes. When the (*s*) axis goes through or near the center of gravity (*CG*) and the turning body remains on the same place, let's call it *spinning*. Let's define *rolling,* when the (*r*) axis passes through or near the body's center of gravity and the rotating object, like on a car wheel, and can also move over a straight line along a surface (or road). When the rotating (*w*) axis is moving inconsistently or unsteadily, let's name this rotation *wobbling*.

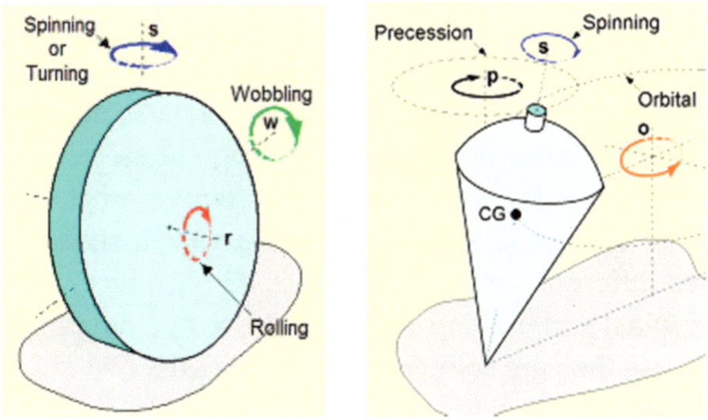

Figure A10. Rotational: movements and axes

When the spinning axis appears to be fixed on a spot and another point on it is moving about another (*p*) axis like the higher spot of a top does, this turning is known as *precession*. It is also known as *Precessional motion*. This movement resembles the conical shape that a fairy or a magician does by moving the wand. When the body's center of gravity, spinning or not, is moving around an (*o*) axis outside the body, we refer to this rotating movement as *orbital*. When a body can simultaneously rotate about two different axes going through its center of mass, for instance *y* and *z* axes in Figure A11, to avoid confusion, let's call spinning to the rotation about the natural rolling (*z*) axis, and *turning* to the other one.

Momentum is the effect (energy) that forces a body or object to keep its velocity and keep moving in the direction of its dominant velocity. If this velocity is linear, the object is forced to keep moving on a straight line, and when its velocity is rotational or angular, it is forced to be on a plane perpendicular to its spinning axis.

Once we get confident and learn how to keep the equilibrium on a bicycle or motorcycle, we are certain that we are not going to fall down anymore during a ride, because the angular momentum (rotation) of the wheels keeps us upright. Best expressed, the *keep in the same plane effect* (KSPE) conspires to have us on the same plane (vertical, or slant if we are taken a curved road) as the wheels of the moving vehicle are. On further paragraphs, let's consider the angular momentum and its KSPE as one and the same. It is easier to move a spinning object along one of the three main (x, y, and z) axes than turning it about any of the other two different to its spinning axis. For instance, it is hard to turn the wheel and the ring shown in Figure A11 about the *x* and *y* axes because they are spinning about the *z* axis.

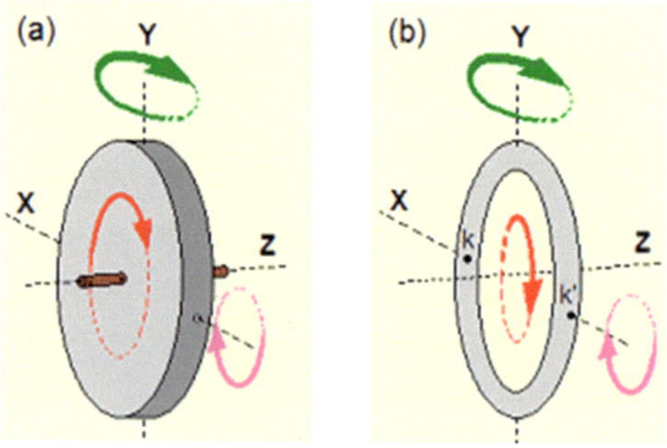

Figure A11. Disk and ring

A *moment of inertia* (I) is a (dynamic) property that measures the magnitude of the tendency of a moving body to stay in its line or plane of movement. It is also defined as a kind of (static) property that measures the resistance of the object to keep its actual shape or external form when its mass is under a deforming action. It is referred to an axis that can be located outside the body. It varies with the shape of the body and how its mass is distributed about an axis or how far the mass is from this axis. The closer the mass is about the axis, the lower the moment of inertia. For instance, on the disk of Figure A11a, the moment of inertia is (half) lower about the x and y axes than about the z axis. This z axis only touches few mass of the disk, or only its thickness, than x and y axes that each touches the disk along one diameter.

Similarly, on a ring its moment of inertia is greater when it is on regard the (z) axis that passes through the ring's center without touching its body than about x and y axes that each touch two parts of the ring. See Figure A11b. On regard to the z axis, almost all the points of the ring are at the same (radius) distance from its geometric center, while on regard to the vertical (y) axis only two (k and k') spots are at the radius distance from this axis, and all other points are closer or touching the ring. See Section *Moment of inertia of some round bodies* for details about this property.

It is worthy to notice that the angular momentum and the moment of inertia are intimately linked. In math terms, the *angular momentum* is defined as the product of the rotational velocity and the moment of inertia, being both referred to and about the same axis. When one of these variables (speed and moment of inertia) increases the other decreases if the angular momentum is pretty constant. A figure skater increases his/her rotational velocity by decreasing his/her moment of inertia by moving arms and legs close to his/her body, and when the arms and legs are moved outward, the moment of inertia is increased and his/her rotational velocity decreases.

Part B *Ro*tating bodies chasing dynamic equilibrium

Don't think, try the experiments.
John Hunter (1728-1793)

On the sections of this part we are going to study the behavior and some characteristics of rotating objects having a mobile balancing mass and a spinning axis of *free* mobility. We also will observe that these bodies tend to reach a "dynamic quasi-equilibrium" state at which they can be still spinning when they sense a disturbing action acting over them. All the cases below deal in some way with the well-known angular momentum phenomenon. Some of the experiments summarized in this section are exposed in section *Homemade experiments in details*. The core of the cases exposed in this section is centered in two physical *imperfections* common to all real bodies: mass distribution and not perfectly symmetrical shape.

The hanging rotating ring

In this experiment, it is possible to observe that a spinning object will move itself or part of itself to different positions where its level of energy gets the steadiest condition and where there is a minimum consumption of energy.

The required components for this simple homemade experiment are a piece of fishing line and a metal ring with three points identified as h, k, and b as is shown in Figure B1a. The ring has to be tied and hanged from a fishing line fixed to an elevated horizontal plane. Keeping the fishing line extended, turn the ring several times. This creates and accumulates a torsion or torque in the fishing line that "pushes" horizontally on the opposite direction of the just applied turning. When the hanged key ring is set free it starts turning about an axis coinciding with the vertical fishing line, and on the same direction of its pushing torque. Again, see Figure B1a.

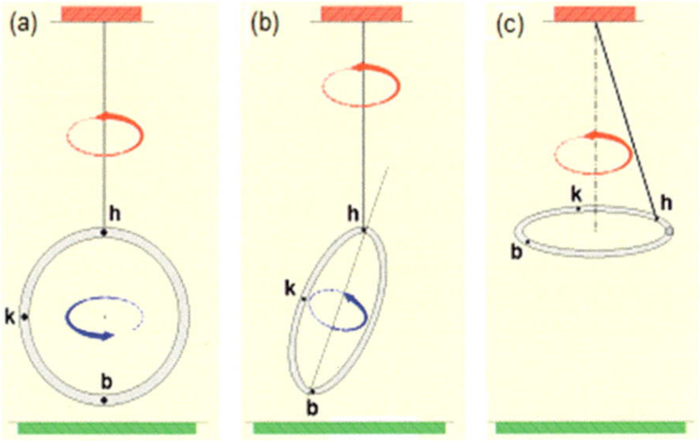

Figure B1. Spinning hanging ring

Let's suppose that the applied torque generates an angular momentum and depending on its intensity can make the ring get and keep for a while revolving and accelerating on the same vertical position. When it starts spinning about the vertical (fishing line) axis, its point k and its mirror one have more kinetic energy than the any other of its points, and its point h has greater potential energy than the bottom point b. These differences on level of energies make the ring search for a steadier position. It slants its rotational axis a few degrees to minimize both the difference of potential energy between the high and bottom spot, and the difference of kinetic energy among the three points. See Figure B1b. However, this position is still unsteady.

If the momentum is strong enough, natural forces will displace the flexible hanging element from the vertical, and transport the ring to the horizontal position where it switches to start spinning about its imaginary and vertical roll-spin axis. See Figure B1c. On this configuration, the steadiest position, all the spots of the ring have the same level of potential and kinetic energies.

Once it is set free, the ring accelerates until all the accumulated twist in the fishing line is released. After that moment and due to its inertial rotation or angular momentum, it keeps spinning and generating in the fishing line a torque that opposes its turning direction. Gradually, this torque gets bigger, and the ring starts descending and its rotation gets slow until it stops and reaches the vertical position again. Then the spinning and the whole process re-starts again but now in the opposite direction. This back and forth is not perennial due to the friction acting over the moving ring, and because there is some loss on the (resilience) capacity of the fishing line to accumulate and release torsion.

When the torque is very strong, the key ring will reach the horizontal position even if it has uniform density, unique cross section area, and is aligned with the true vertical. Why does the ring move to the horizontal position? It is chasing a steadier condition where it is at a minimum consumption of energy and where this energy, is also in harmony with all its evident forms (potential, kinetic, etc.).

For a small interval of time during which the angular momentum of the ring is constant, the ring reaches its maximum rotational velocity (kinetic energy) on the vertical position, (Figure B1a), where its moment of inertia is less than the one on its horizontal position. From the point of view of energy, on the horizontal position (Figure B1c) its rotational velocity is lower because all the points of the ring have more potential energy that has been interchanged by a decrease of its kinetic energy.

For a short period of time, the ring energy on any position is almost the same, it moves itself to the horizontal position just chasing more stability. The ring will also chase this position when it senses any small imperfection like a spot of different density, or a tiny deviation from the true vertical. Notice that in addition to the upward force of the fishing line, there is another upward vertical one opposed to gravity force that lifts the ring.

What is described on the above paragraphs can be easily proved in a homemade experiment, but the “size” of the ring and the flexible fishing line should be on the same order of magnitude. An excessive torsion on a short and strong fishing line could decrease (see the rotational squeezing effect) excessively its length and create an up tension much greater than the ring weight, making its pull up so strong that the ring at the moment of its release instead of rotating, will be shot up towards the hanging point. A fishing line of 2.7 Kg (6 Lb) and 0.22 mm (0.009") diameter works for the split key ring, but one of 22.7 Kg (50 Lb) and 0.71 mm (0.28") diameter is too strong.

The experiment of the next section is also with a ring that is spinning on a flat glass, and about one of its vertical diameter.

The spinning split key ring

The behavior of a split key ring spinning around a vertical primary axis will help you to understand some characteristics of solids on revolution, when they are not symmetrical and/or their mass is not uniform as is the case of our planet Earth. The observations seen in this experiment helped to infer that the lateral or orbital spinning of a rotating object is promoted by an energy imbalance chasing stability and a dynamic equilibrium.

Figure B2. A split key ring spinning

The split key ring has a uniform weight by unit length, except by its *mouth* (place through where the keys are put into the ring) that is formed by only one ring while the rest of the key ring, is by two. The mouth spot is lighter and weighs half grams by millimeter less than any other ring zone having its same length. Figure B2 shows a spinning key ring having a lead sliding over a taut string coinciding and connected to the ends of its gravity diameter. That for observation purposes is identified and limited by the red painted mouth (*m*) and its diametrically opposed point, or black painted *contra-mouth* spot. A big token or coin with a relative big orifice near its edge to make it asymmetrical can be used instead of the split key ring to carry on the experiment.

The ring during the spinning describes the same typical movements of a top (see Figure A10) and it also adopted the following six (6) typical positions or stages:

- 1st. The string is at or close to the horizontal position, and the bead is against the ring.
- 2nd. The heavy or contra-mouth spot goes up at or close the spinning axis, and it remains at the top semicircle until the fifth stage. The bead either remains against the ring or close to it in the top or bottom semicircle of the ring.
- 3rd. The bead moves to the bottom end of the string if it is not already against it.
- 4th. The gravity diameter moves to an intermediate position between the vertical and the horizontal.
- 5th. the ring movements are unstable but it still keeps the upright position.
- 6th. No more spinning. The ring starts falling and rebounding until all movements end.

The ring rotational speed is maximum, just at the moment (time zero) when it is spun out and starts gradually decreasing to zero as the time passes by. This decreasing in spinning and the mass imbalance promote the split key ring stages. In other words, if the spinning were constant only one stage would be active.

Imagine that we horizontally cut the key (double) ring into pieces having the same length than the mouth's, and we group them in pairs, and keep them spinning together at rotational speed of the split key ring. Each member of a pair should have two pieces of the double-ring as the one shown in Figure B3a. All of them are identical except for the unique couple having the mouth as one of their components. See Figure B3b. On each of these pairs there are three kind of energies: the weight (*W*) or potential energy, rotational kinetic energy (*K*), and the angular momentum (*M*). All of them are proportional to the mass of the pair. There is a fourth energy, the friction action, whose main function is to decrease the rotational speed.

To visualize the effect of these energies on the behavior of the split key ring, in Figure B3a, force yourself to imagine that on each couple there are acting simultaneously the following three imaginary things or devices: one seesaw plank where the members riding on its ends are the potential energies (*W*), a tug-of war contest in which the teams pulling again each other are the kinetic energies (*K*), and a monoplane or boat propeller whose two spinning blades are the couple of angular momentum (*M*) of a key ring pair pieces.

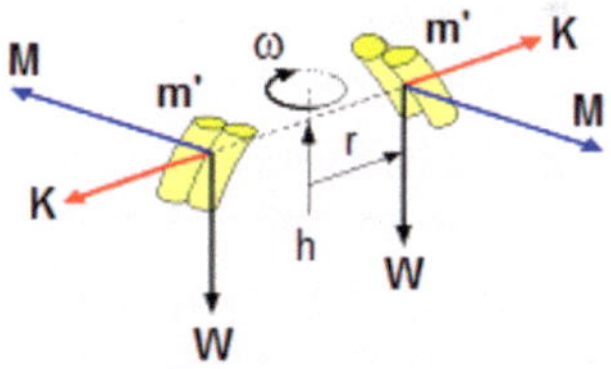

Figure B3a. Energies with equal magnitudes

On each member of a pair at the same height (*h*), the seesaw effect or torque of the weight (*W*) acting downward at each side of the rotational axis, cancels each other out. On the other hand, the rotational kinetic energy (*K)* on each sector are equal and are acting over the same horizontal straight line (they are collinear), but acting in opposite directions, therefore they also neutralize each other. The magnitude of their angular momentums (*M*) are also equal, but they have opposite directions acting over two different horizontal parallel lines. However, their pushing torque about *S* axis is in the same turning direction, so they add their effects and push to keep the ring spinning on the same revolving "direction". That is also true when the mouth is part of a "pair" (Figure B3b). Thus, what keeps the split key ring on the upright position and spinning is just its angular momentum (the pairs of energy not auto-neutralizing themselves).

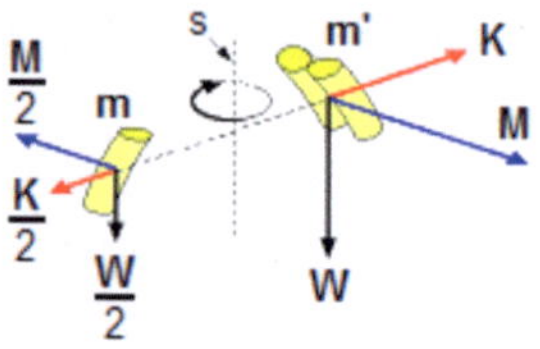

Figure B3b. Energies with unequal magnitudes

When the lighter mouth sector (*m*) is part of a pair of pieces as it shown in Figure B3b, it creates at that level an imbalance, specifically the energies (potential, kinetic, and angular momentum) on the mirror sector (*m'*) are two times greater than on the mouth (*m*) side. At this instant, both the rotational kinetic energy and the potential energy push, (down and to the right) trying to make the ring roll to its heavier side. Once the ring turns (180 degrees) half a revolution, the mouth switches position with its mirror sector, switching the rolling effect to the opposite direction too. This effect that should be at different height is the one that promotes the lateral or orbital movement of the ring. This switching effect, a very unstable condition, (up and/or down, to the right and/or the left) will continuously happen while the ring is vertically spinning. Consequently, the ring mouth should move itself to a more stable condition, namely to a location where the imbalance is at its minimum.

The position-switching of angular momentum of the mouth does not have an unstable effect on the ring because both components of this pair keep pushing on the same turning direction. At the beginning they transport the mouth to, or near the horizontal diameter where the potential energy is less disturbing, because it is almost equal on either side of the ring.

Contrary to a mass whose keep moving effect is stronger when it is far from the spinning axis, the moment of inertia of the missing mass[8] of the mouth is maximum when it is close to the spinning axis or when it is either at top or bottom the vertical diameter of the ring, see *Moment of inertia of some round bodies* for details. Knowing that, on which of these positions should the mouth be so that the split key ring gets its maximum stability? Nature moves the lighter sector, the mouth, to the bottom from the second to the fourth stage of the spinning key ring. Why? This fact is contrary to the one that a seesaw does: it always takes the end with the heavier weight to the bottom and the lighter up.

Nature would do that too if the split key ring were "always" spinning about a true vertical axis (on many tries it was observed just twice). However, actually the top of the ring is almost always a bit slant and on precessional rotation. This movement is more stable if all the mass describing the precession is uniform on all the precessional trajectory, and it happens when the contra-mouth is on the top section

This effect can be explained considering the precessional rotation, and recalling that what tries to keep the angular speed constant is the angular momentum. As more mass is far from the spinning axis, the stronger the effect of the precession. To visualize this, imagine (or try) making a conical movement (precessional spinning shape) with a magic wand having one end heavier than the other (like a hammer), and do it first by grabbing it by one end, and then by the other. By what end should we grab the wand to make it more difficult to stop once it is moving? By the lighter end to keep the heavier part far from you hand, do you agree? Hence, the split key ring kept its mouth on its bottom section just for stability. It takes precedence over its potential energy that is a bit less if the mouth were on the top.

---

[8] The mass needed to complete two rings in the mouth sector.

On the other hand, the unstable switching and rolling effect caused by the mouth-potential energy pair described previously is minimized when the unbalancing potential energy is minimum or when the mouth is in the bottom section of the split key ring.

This ring spinning on a surface and the one hanged spinning ring behave in a similar way. They move their mass to a position where the consumption of energy and stability or steadiness both reach a minimum. While the hanging ring generates an upward lifting force; the surface-spinning ring, a *torsional force* that moves the mouth over its perimeter. According to the results of this experiment it is possible to conclude that the lateral or orbital movement of the spinning key ring is generated by an imbalance of energy.

The next experiment with a toy gyroscope is a kind of hybrid of the experiments of the two previous sections, but a disk is used instead of rings. It is similar to the test with the hanging ring because the turning torque in both bodies is applied through a hanging fishing line. It is spinning as the split key ring, but the disk is spinning about its rolling axis instead of about an axis going through one of its diameters.

The spinning and rotating toy gyroscope

In this experiment a centered hanged and spinning toy gyroscope is disturbed by external actions (torque and/or a vertical force). That, according to the observations, generates two lateral forces and movements in the toy: one parallel and the other perpendicular to the spinning plane. The knowledge gained in the experiments, will allow us to predict the direction of these natural movements and forces before they begin. It was also discovered that a spinning object can simultaneously have two instantaneous centers of gravity. For details see section *The spinning toy gyroscope*.

Centered hanged means that when the toy gyroscope is hanged and gets balanced, its static wheel is on an upright position, and its center of mass is in the midpoint of the length of its horizontal axle and in the middle of the wheel.

The toy keeps spinning in the same hanging position when it is hung from a fishing line free of any torque. However, when a disturbing torsion in the fishing line tries to turn and rotate the toy gyroscope about the vertical (y) axis, initially its strong KSPE effect of its spinning wheel does not allow this action. However, as the disturbing torsion keeps *pushing*[9], the balance is broken and the toy starts chasing equilibrium. It begins an incipient lateral or orbital movement and lifts a bit the bottom of its wheel as it is shown in Figure B4a. The two quite different movements suggest that two forces are generated, and/or that they may occur in two different spots.

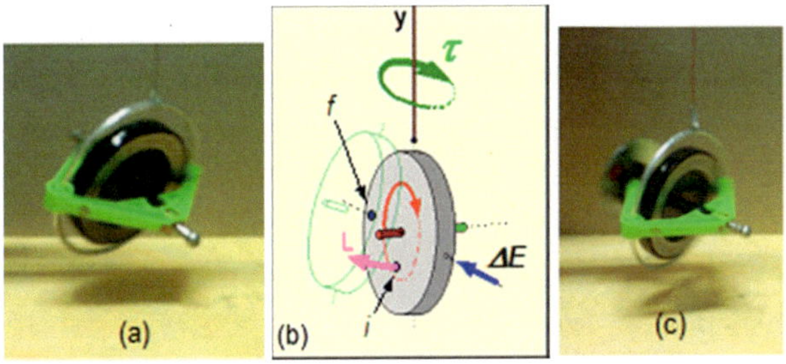

Figure B4. The centered hanging toy gyroscope

The magnitude of the kinetic energy is the same in all its peripheral points. On the other hand, the potential energy on the wheel, is maximum on the top and minimum on the bottom, and has the same medium value in both ends of the horizontal diameter. When the effect of the peripheral potential and kinetic energies are added in the (four) ends of the vertical and horizontal diameters, it is found that the strongest and the weakest accumulation of these energies are in the horizontal diameter of the wheel. The greatest concentration of energy in the horizontal diameter was defined as the *excess of energy* (ΔE) or latent action

[9] This pushing is actually a turning effect applied on the entire toy.

whose direction aims towards the weakest spot. To counteract this excess of energy, an opposite one was defined as the *reacting energy* (ΩE). Using these four accumulations of energy, were also found two instantaneous (*i* and *f*) centers of gravity shown in Figure B4b. In these spots are where the lateral-orbital force and lateral-lifting force act.

The lateral force parallel to the wheel's plane that creates the orbital velocity, is governed by the spinning direction, and always points the same as the *excess of energy* in the wheel, or from its geometric center on axis (axle) towards the *f* spot. This lateral movement should be on a straight line but because the toy gyroscope is tied to the fishing line, it is forced to start orbiting around the vertical axis going through the hanging point.

It is worthy to clarify that the torque (τ) in the fishing line used on the experiments is variable, alternating, and cyclical. It cycles in two stages: in the first it releases and transfers its effect to the ring from a maximum value to zero, and in the second it accumulates from zero to a new maximum value the torque received from the toys' ring. It means that once the lateral movement has started, it and the moved up axle end are kept in this particular direction until the toy regains the lowest or switching *position.* That is reached when the spinning toy is free of any turning movement and the torque in the fishing line has arrived at a maximum value. In this static position, the spinning wheel interchanges both its orbital direction and the lifting of one of its axle ends. After the switching, the process of releasing-accumulating energy in the fishing line re-starts.

During the releasing stage, the toy gradually absorbs and converts the torque into orbital momentum that increases both the orbital velocity and the height of the wheel to certain maximum values. That are reached every time the fishing line is free of torque. After that, these velocity and height gradually decrease while the fishing line is accumulating the opposing new torsion. For being associated to both the spinning and the fishing line torsion directions, when one of these parameters change, the direction of the lateral movement and the lifting of the axle's end change accordingly.

When at the moment of hanging the spinning wheel, two disturbing effects (a fishing line torque, a force and an artificial or elevation/depression of an end of the wheel axle) are applied simultaneously, but acting in different directions, the system or nature may start orbiting in the direction of the stronger effect. If the torque is the strongest, it will first or simultaneously switch down/up the axle end that was artificially elevated/depressed. This amazing switching will occur whether or not there is a load added to an end of the axle (Figure B4c).

According to this experiment, if a wheel (or object) is spinning in a fixed location, and it starts a lateral (orbital) movement, there shall be an external stimulus promoting such movement.

The offset-hanging spinning-turning wheel

To validate the results obtained with the toy gyroscope, a similar experiment can be (and was) conducted with an offset hanging bicycle wheel, but in this case, the disturbing torque is constant and tries to turn the wheel about a horizontal axis instead of one vertical. The main purposes of the experiment were: first, confirm that the external movements (including the orbital) of a spinning object, emerge as a reaction to actions breaking the equilibrium of the spinning wheel; second, validate that this orbiting is in the direction of the excess of energy.

Figure B5a shows the bicycle wheel spinning while hanging from a two-leg hook located at one end of its axle that is extended by a support. This custom made hook acts as a fulcrum, a kind of hinge that allows a freely lean down of the wheel, that when is not spinning will abruptly incline, and align vertically with the hanging point *A*, the *hanging* (*O*) *point,* support's centerline, the axel of the wheel, and its center of gravity (P) point. It also means that none of the pulling up forces conveyed throughout the hook to the hanging point is transmitted to or reaches the wheel. Three hanging arrangements were used: a fishing line as the one shown in Figure B5a, a spring as is shown in Figure B5b, and the hook attached directly to a sturdy frame. Three parallel perforations were drilled throughout a diameter of the support to create the three tested offset (*OP*) distances.

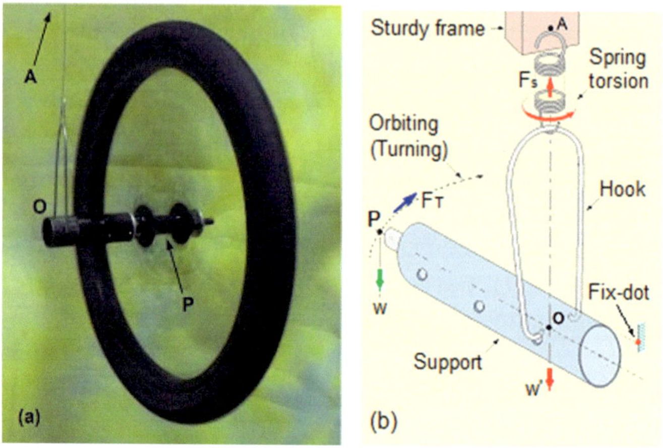

Figure B5. Offset-hanging wheel: a) spinning, b) main forces

On the test with the fishing line, once the spinning wheel is released in the upright orientation, it starts orbiting for a while around an axis passing through the *hanging (A) point*, describing almost perfect horizontal circles. The slower the spinning, the faster the descending or tilting rate. At very low speed, the hanging *O* point and the fishing line do not remain at one location anymore and starts getting circular (orbital) movements about an imaginary vertical axis going through the hanging (A) point.

For the test with the hook attached to a spring, and during the spinning test, the *hanging (O) point* is free to move up and down. However, during most of the orbiting-descending process, this point remained in a fixed spot over a horizontal plane. That is possible because the angular moment of the wheel creates two vertical forces of opposite directions, specifically one to lift up the net orbiting weight (*w*) acting on the *center of gravity* spot (*P*), and the other (*w'*) equal to the overall weight acting on the hanging point (*O*) to pull down the spring and keep it tense, as it is shown in Figure B5b. Therefore, in the vertical plane, the spinning wheel (its angular momentum) simultaneously generates two non-collinear and parallel vertical forces having different magnitude and opposite direction. On the other hand, for having one (A) end fixed against a sturdy frame, the spring can withstand torsion. The magnitude of this torsion is proportional to the turning angle that the spring twisted or turned by the horizontal traveling (orbiting) of the wheel. This resistant twist of the spring observed in the test is proof of the existence of both a horizontal lateral (orbiting) force (F_T) that induces the orbital movement, and the horizontal torque it creates over the spring. See Figure B5b. This shows a similitude with the spinning split key ring!

Experiments were also done attaching the hook directly to a sturdy frame. For the medium offset distance, a stronger hook made of a steel bar of 4.76 mm (3/16") was elastically deformed and twisted. Due to the magnitude of this deformation, these experiments were not attempted for the greatest offset. Similar results but not as dramatic as the ones obtained for a bicycle wheel, were gotten from a toy gyroscope when it was both offset and centrally hanged from a metal wire of 1.6 mm (1/16") that offers to the toy an almost infinite resistance to the torsion. This test also confirms the existence of the strong force (Ft) inducing the lateral movement.

When there is no offset, or when the hanging point is aligned with the center of gravity (point P) of the upright spinning wheel, there is not a steady turning movement in any direction. Instead, the wheel either just turns a few degrees to one side and then returns, and then keeps going back and forth between these ends, or the turn just reaches and remains on one end until the spinning is very slow or ceases completely. This test also confirms that the lateral movement is a counter action to the disturbance caused by the offset torque.

From several tests, it was observed that the orbiting direction is always in the movement direction of the bottom spot of the upright wheel. That matches with the behavior of the spinning toy gyroscope. So, on the bicycle wheel there are also present the instantaneous *i* and *f* spots, and the excess energy and its reacting energy.

From the experiments with the toy gyroscope and the spinning wheel, one might conclude that the external movements (orbital, wobbling, lifting, etc.) of a spinning object emerge in opposition to disturbing actions acting upon the object. Is the orbital movement of the Earth promoted by an energy unbalance? Can the physical medium, the support, interconnecting the hanging point and the spinning object be necessary if the tilting effect (gravity) disappears or is very small? Can the resistance to turning disappear completely? Some answers to these difficult questions are revealed in the next section. This section covers a case where gravity is aligned with the orbiting movement instead of acting perpendicular to the orbital trajectory as occurred in the experiments with the toy gyroscope and the bicycle wheel.

The skipping cylindrical bomb

This section is a summary of a PBS documentary Titled "Bombing Hitler's Dams" (Duncan 2012 WGBH). During World War II, the British planned to bomb Hitler's two gigantic dams that were protected with large reaching artillery that made it too risky and almost impossible to destroy, even from any close point, or by airplanes dropping torpedoes that could be stopped by strong metallic nets that were deployed near to the dams. The British created a cylindrical spinning bomb to be dropped by an airplane flying at low altitude from the water, and a far distance from the target dams that would be reached by the bomb skipping over the water. See Figure B6a. During initial testing it was found that an impact against the top of the dam would only dent its wall, or would make a hole on it with no destroying effect. This effect is similar to the one gotten during an explosion when the explosive is located on the surface of a thick metal plate, and covered with just a plastic cup full of water located upside down over it. The explosion on this arrangement really damaged the plate in comparison to almost no damage when nothing covered the explosive during the explosion.

To destroy the dams, the explosion should be very close to the wall and near the dam foundations. So, the bombs not only needed to reach the dam, but also remain as close as possible to the dam wall while diving down. How could the bomb be kept close to the wall?

On initial tests with small balls, they were shot almost parallel to water level, and spinning frontwards[10]. As soon as they touched the water, they tried to start rolling but finding no friction on the water, they kept rotating on one spot without jumping. This rotating action digs a cavity in the front of a ball promoting its diving. Once the spinning direction was reversed, the balls started skipping. Due to failures and destruction of spherical prototype bombs, and for easy construction, its shape was changed to cylindrical.

Upon touching the water, the cylindrical and back spinning bomb rapidly accumulated water near its front, and above the water normal level. This impacted region acted like a hard surface, see the box area in Figure B6b, against which the cylinder collided to. The water accumulation, its dispersion and splashing occurred in a very short time, but enough to make the bomb rebound.

In addition to the rebound effect, when the back-spinning bomb touched the water, its bottom side lost speed due to the friction, while its top section, free to rotate, kept its initial rotation that created a speed imbalance at these two sections, or points A and B of the bomb. See Figure B6b. This imbalance, instantaneously made the cylinder rotational axis move to a point o'[11] located where imaginary lines p-q and A-B intersect. Also two

---

[10] Frontward spinning means that the bomb top section is moving the same as the airplane's moving direction.

[11] For details about this instantaneous spot, see *Energy, mass, and center of masses* and Figure A5a.

instantaneous moments are generated around such points with levers are *a* and *b*, *b* being greater than *a*. Due to this difference on levers, the "seesaw" moment on A side, aiming slanted down, is lower than the another on side B, aiming slanted up, so the moments net effect is aiming up, helping the bomb jump.

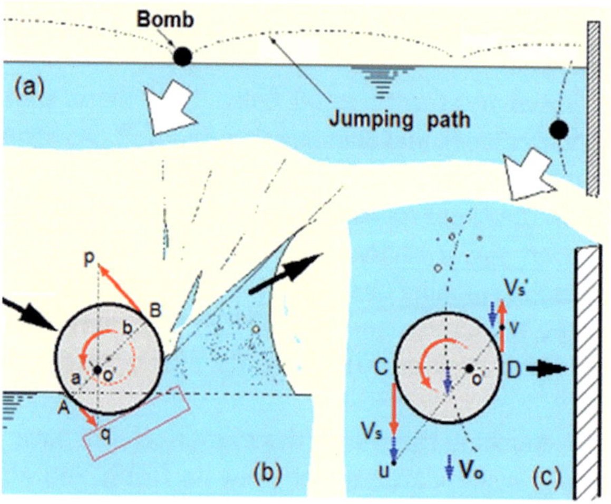

Figure B6. Phases of the bomb skipping

After hitting the wall of the dam, bouncing away from it, and falling through the water, all the cylindrical surface of the cylinder acquires a tangential velocity *Vs* due to the spinning. All points of the cylinder also acquire a speed *Vo* downwards. This speed adds to the tangential speed *Vs* aiming downwards on its side C far from the wall and, it subtracts from the tangential speed V_S' (same magnitude as *Vs*) aiming upwards on its side D close to the wall. See Figure B6c. That addition/subtraction makes the superficial velocity on side C be greater than on side D. This speed difference creates an imbalance which momentarily moves the rotation axis from its geometry center to point o' that is located where the imaginary line u-v is intersected by the imaginary line C-D. In regaining its rotating balance, the unbalanced energy tries to move the rotation axis to such point o', dragging and moving the cylinder toward the wall. This dragging effect (force), perpendicular to the travel speed, is known as the Magnus Effect (Wikipedia, Magnus Effect). It will keep moving the cylindrical bomb toward the wall until it hits the wall and rebounds back, or until the spinning and/or falling stops.

There are four important facts hidden and embedded on the traveling of the spinning skipping bomb that must be highlighted. The first is that the net effect of spinning plus straight traveling will generate lineal velocities of different directions and magnitude on two opposite spots of a submerged cylindrical object, as is seen on sketches (b) and (c) of Figure B6. The second is due to the difference of superficial velocities; the center of gravity of the spinning object (or the entire object) will move mainly towards the direction and closer to the spot out of the two with the lowest lineal speed. Third, submerged, and surrounded by a fluid or energy, the moving and spinning object will create a curve path, orbit, in which concavity aims towards the same side where the spot of lower (energy) velocity is located, as shown in Figure B6c. From this fact, the higher the spinning, the higher the path curvature, and the higher the straight-line velocity, the flatter the path curvature. In other words, at no spinning the moving path should be a straight line.

It is amazing that the skipping bomb was conceived and developed from an ordinary fact well known to kids and youngsters living in the vicinity of a brook, pool, or lake. Do you remember throwing a flat stone on the water surface and watching it jump? The goal was to see who can make it jump more times, faster, and farther. Although skipping stones in water might be considered a childlike activity, with these challenges, one learns hands-on that the stone should be spinning as much as possible, and that it should be thrown almost parallel to the water's surface, two of the most important facts needed to make the stone and ... the bomb jump. Details about the actual development of the skipping bomb can be watched at the fascinating PBS documentary "Bombing

Hitler's Dams" (Duncan 2012 WGBH). I deeply hope the reader will be compelled to watch and enjoy this must-watch documentary.

A throwing stone and a typical projectile describe a parabolic-curved path because of gravity acting downwards and on certain positive angle over their traveling trajectory. On the submerged and diving, spinning bomb, gravity does not have a direct influence in the curvature of its path because it is moving on the same downward direction as (or parallel to) gravity. Its curved diving path is generated by the combination of two effects, namely spinning and a small energetic unbalance, both conspiring against traveling through a straight line by creating a horizontal force that is perpendicular to gravity. Another issue that is worthy of mention is that no physical media, like a fishing line used on the toy gyroscope or the extended support of the offset hanging wheel, is needed to promote the curved path.

Balls spinning about horizontal axes

Suspended balls spinning about horizontal axes should have a behavior similar to the Earth's. That could be also similar to the one of the split key ring spinning on one of its vertical diameters. The imbalance along the gravity diameter can induce rotation on the suspended balls. Corroborating such hypothesis is the main purpose of this experiment.

When placed on top of a vertical stream of air, a ball starts "floating" in front of the pipe outlet rim, then some time later, it is forced to move toward one side of the pipe edge where the air is partially blocked and deflected to the diametrically opposite side. In consequence, through this "open" section there is an overflow of air (or energy) that for the air-ball friction effect, drags up this side of the ball promoting its horizontal spinning. See Figure B7a.

If on a uniform and homogeneous flow of air were placed a ball, having uniform mass, which geometric center and center of gravity coincide, its spinning would reach a dynamic and stable equilibrium. That is, it will spin at a constant rotational speed with the rotational axis on one and only one horizontal location. Any point of it, like the *H* point shown in Figure B7c, would keep rotating indefinitely on the same vertical plane at the same spinning speed without any other velocity.

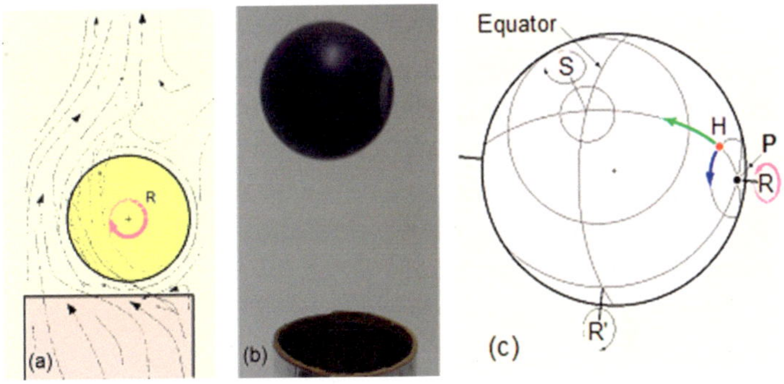

Figure B7. Balls floating, and spinning about horizontal axes

However, if the geometric center and center of gravity do not coincide, the ball could not get to a stable equilibrium because the *H* point in addition to spinning also has another velocity along its spinning axis. Gaining the equilibrium or looking for a stable condition, the spot while spinning has two options to (1) move toward the rotating axis and reach the P (pole) point, or (2) move toward the equator. For some spinning values, this spot kept moving on this direction and surpassed the vertical equator, switching place (poles) with the lighter spot. For other spinning speed values, the heavy point keeps moving most of the time between a pole and the equator without reaching either of them. Figure B7b is a snapshot of an actual ball in this situation.

A heavy section or spot clearly identified on a ball that is spinning around a horizontal axis, has a behavior like the split key ring. See Section *The spinning split key ring*. At high spinning speed, the heavy spot spins far away from the rotational axis (that is equivalent to the first stage of the split key ring). At an intermediate spinning speed, the heavy spots reach a position near to the rotational axis. That condition is equivalent to the second and fourth split key ring stages. The "heavy" and "light" spots almost never get aligned with the rotational axis. This case resembles the fifth stage of the split key ring rotation. It is worthy to notice that we are certain the split key ring has a lighter spot (the mouth) located on its external surface.

The balls on the energized (air) streams point out that the interchange of flux of energy (quantity of energy by unit of area) between two sources or bodies, shall occur gradually and at levels of the same order of magnitude, or that one with the most high level will entirely absorb or minimize the one with the low level, or will blast it out. The Earth is not incinerated and/or blasted away by the Sun because we get from it just enough energy to keep the Earth moving.

Balls revolving about an almost vertical axis

Most of the "big and heavy" balls just float with none or few spinning about an almost vertical *S* axis following a wavy and wobbling orbital *P* path as shown in Figure B8a. Others, like the one ("earth ball") in Figure B8b also follow a path *P* with nutation[12] movements. Some of them, as soon as they reach a high rotating speed, switch their rotation axis to a horizontal position. On the contrary, the red ball (Figure B8c) almost never rotated on a horizontal axis. The red ball not only is heavier than the "earth" ball but also its heavy spot to overall weight ratio is much greater.

The red, hollow, and bigger ball has light and heavy spots that resemble the Earth poles, they are part of the ball hemisphere sides joined by a seam also resembling the equator. The heavy spot of it, or one of its "poles," is clearly identified by an accumulation of material forming a small and solid spherical cap. See the bottom of the red ball in Figure B8c. When their heavy spots were on the top or above the equator, the two quite different balls spun at almost the same rotational speed (35 rpm)! When this spot was on the bottom or below the equator, their rotational speed rose, but less on the heavier red ball than on the lighter "earth" ball (52 rpm

[12] Wavy and oscillation of a spinning object.

Vs 200 rpm) that sometimes reached values of about 550 rpm. It seems that when the heavy spot is on the top, a condition that happens most of the time, some energy is used to keep this section up. When the heavy spot switches to the bottom, part of that energy is converted into rotation.

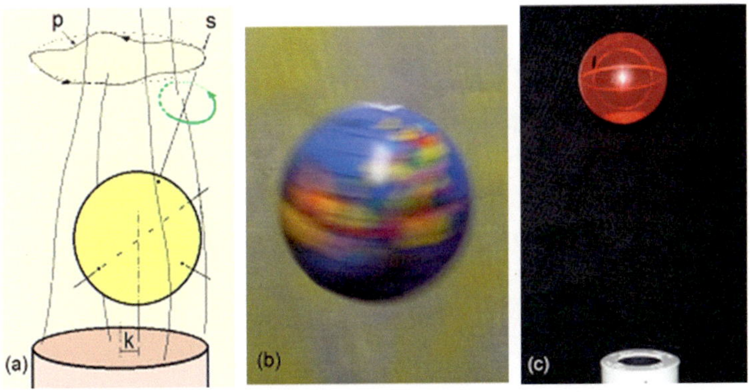

Figure B8. Balls spinning around a "vertical" axis

The movement of the red ball spinning about a "vertical" axis resembles the movement of the Moon around Earth. The Moon oscillates, keeping the same face toward Earth at all times while orbiting and wobbling (going up and down) with its axis tilted a few grades to the orbit path (Wikipedia, Orbit of The Moon - Libration, 1st paragraph). A notable difference is that a ball spins about an axis passing through or near the geometric center of its body, while it is believed that the Moon does not spin about its own axis, but orbits following the Earth-Moon common rotation axis.

In the experiments with balls spinning while suspended in a vertical stream of air, it was observed that the balls' rotational axes were a bit slant to the gravity diameter interconnecting their geometric center and their center of mass. These centers did not coincide on any of the eccentric[13] spinning balls. When these centers were close (low eccentricity), the balls' spinning axes were almost always close to the horizontal and were rotating at higher velocity than those balls of higher eccentricity whose rotational axes were close to the vertical. The spinning axes' orientation and the balls' rotational speeds suggest that the rotation is induced by the balls eccentricity.

Group of rotating objects

The two experiments having a free rotating axis described in this section show how a body out of dynamic balance will rearrange its configuration chasing a dynamic equilibrium, as it also did the hanging ring and the spinning ring.

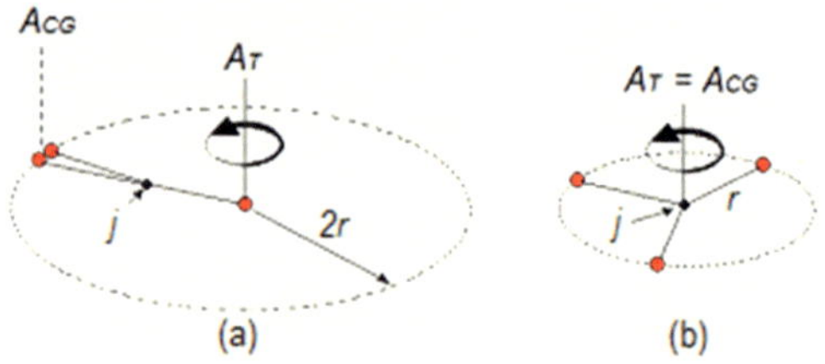

Figure B9. Three balls with free rotating axis

The first experiment consisting of three balls having the same weight, each connected to ropes with the same length and joined on a common point *J*. It is a typical South American device called *Bola*. One ball is used to spin up the other two as it is shown in Figure B9a. Then, they are tossed to a target on an unbalanced spinning: two balls revolving against and about one ball. Once on free movement and after a while, they rearrange themselves into a more stable configuration in which the ropes of each ball is 120 degrees apart from each other. See Figure B9b. In this new

---

[13] In this text, eccentricity means that the center of mass does not match with the geometric center.

configuration, the balls spin at higher rotational speed because they reduce their moment of inertia by getting closer to the spinning axis.

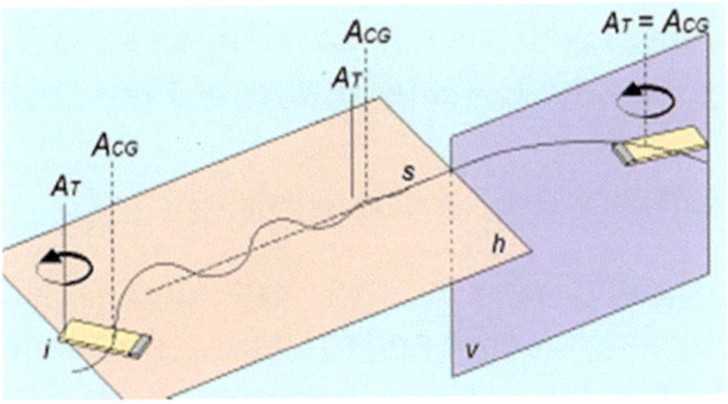

Figure B10. Flat object with free rotating axis

During its transition stage from the initial condition (*i*) to the condition (*s*) when the body starts moving over a straight line, the center of gravity of *Bola* may describe a path like the one shown in Figure B10. To describe that trajectory, let's suppose that a flat, rectangular, and asymmetrical object[14] is tossed with rotating speed across a horizontal plane *h*. At the throwing moment (*i*), the object will start rotating around a temporary A_T axis located at the releasing point, corner *i* in Figure B10, that is a certain distance from the object center of gravity (*CG*) intersected by the A_{CG} axis. At the beginning, the A_{CG} axis movement will follow an alternate trajectory, and the A_T axis starts "erratically" rotating around the center of gravity, and getting close to it. Once both vertical axes (A_T and A_{CG}) coincide and act as a unique rotating axis, from point *s*, the object linear and rotational energies start moving following a unique descending and steady path on the vertical plane *v*.

The moving path of the two dissimilar objects shown in Figure B9 and Figure B10, describe a similar pattern from the instant (*i*) of tossing them to the moment (*s*) that their initial rotating axes follow and coincide with the center of gravity. So, it is possible that it should exist both a mathematical equation

[14] Like a metal bar attached to an end of a timber board.

describing such path and a physical law that explains in detail the behavior of those movements. That is not an easy task as stated by Stephen Hawking who said "We cannot even solve exactly for the motion of three bodies in Newton's theory of gravity, and the difficulty increases with the number of bodies and the complexity of the theory" (Hawking, A Brief History of Time 186)

A summary about rotating bodies

After dealing with all the experiments looking for an explanation to why the orbit and rotation of our planets have been associated to the law of universal gravitation, I have reached conclusions that I want to share with you. I hope the reader has shared and enjoyed my curiosity in this searching. The conclusions reached are based on experiments with spinning bodies (hanging ring, split key ring, toy gyroscope, bicycle wheel, and small balls), and a documentary about a spinning cylinder. Table C1 summarizes the experiments. The second part of this summary is exposed at the end of Part C.

In the hanging ring, its rotation about a vertical axis is induced by an external and cyclical torque. Once spinning in the vertical orientation, the ring created different levels of potential and kinetic energies on different points of its surface. These different level of energies are unstable and the source of the disturbing effect. Therefore, the ring chased a more stable position moving itself to the horizontal position where all its points had the same level of potential and kinetic energies. To move from one position to the other, the ring generates a force. This force was initially horizontal and then was vertical to move the ring from its vertical and unstable position to its horizontal and more stable position. This hypothesis of the existence of these forces was validated in the experiment with the bicycle wheel.

In the split key ring spinning about an axis in an almost vertical orientation on a flat surface, the disturbing effect was caused by a mass missed on one of its two rings. This ring sector of missed mass moved itself around the ring surface by a tangential force to compensate the changes of type of energies in the ring due to the gradual decreasing of the rotational speed of the ring. The orbiting and precession of the ring were also caused by the missed mass causing imbalances of energy in the ring. The tangential displacement of the missed mass with the rotational speed suggests that there exists a direct relationship between the position of such mass and the spinning speed.

Table C1. Experiments summary– Created forces Vs Imbalance

Feature / Body	Support type	Spinning source	Imbalance		Created Force			
			External	Internal	O	L	E	T
Hanging ring	Hanged	Induced	None	Energy	---	√[2,3]	---	---
Key ring	Flat surface	Given	None	Mass & Energy	---	---	---	√
Toy gyroscope	Center hanged	Given	Vertical torque	Energy	√[2]	√[2,3]	---	---
Bicycle wheel	Offset hanged	Given	Horiz. torque	Energy & force	√[2,3]	---	√[4]	---
Diving cylinder[1]	None	Given	None	Velocity	√[2]	---	---	---
Suspended balls	Air up stream	Induced	None	Mass	---	---	---	√

O= (Orbiting) on spinning plane
L= (lifting) perpendicular to spinning plane
E= External to spinning plane
T= (tangent) Turning other than spinning
(1)- Data from a documentary
(2)- Horizontal
(3)- Upwards
(4)- Downwards

The hanging spinning ring, and the split key ring spinning on a flat surface generated natural forces with different orientations because the effect disturbing their spinning was also different. However, in both experiments, the rings were chasing a steady position where their bodies were more balanced energetically and the consumption of energy was minimal.

In the centered-hanging toy gyroscope whose wheel spins about its rolling axis, the main disturbing effect was an external and cyclical torque that tries to turn the wheel about an axis parallel to and passing throughout its spinning plane. This warping torque magnifies the potential and kinetic energies on the perimeter of the gyroscope's wheel. To counterbalance these imbalance of energies, two perpendicular forces, acting on different planes, appeared. One of them promoted the orbital movement of the gyroscope along the wheel spinning plane while the other inclined it. The orbital force was always perpendicular to gravity, but the inclining or lifting force varied its inclining angle from perpendicular to almost parallel and opposite to gravity.

In the offset hanging bicycle wheel spinning about its (axle) rolling axis, the disturbing external effect was an almost constant torque generated by the wheel hanged from a spot that does not pass through its center of gravity. In other words, the wheel was kept suspended from just one of the ends of its axle. Three natural forces were generated to counterbalance this disturbing torque. One of these forces induced the orbital movement along the wheel spinning plane, and the other two parallel forces of opposite direction were perpendicular to the first one. One of these two parallel forces was upward and kept the wheel in its orbital plane and kept the wheel from falling while orbiting, and the other downward force acted on the hanging spot. By knowing the magnitude of the orbital force, allowed us to validate that the angular momentum of a rotating body (bicycle wheel and toy gyroscope) is stronger than its orbital (linear) momentum.

When the toy gyroscope and the bicycle wheel were centered hanged, they remained spinning in a fixed position when they were dynamically balanced and nothing disturbed them. However, when disturbed by an external effect, they started chasing equilibrium and generated some forces described in the previous paragraph. This chasing action also appeared in the hanging spinning ring due to differences of levels of energy and the spinning split key ring due to the *revolving mass* imbalance. Due to differences of its peripheral velocities, a similar force horizontal and perpendicular to gravity was also observed in a spinning cylinder diving in water. This force creates a horizontal velocity that combined with the falling velocity generated the curved trajectory of the descending diving cylinder. Therefore, according to the results of these "five" experiments, one can infer that when one spinning object, for whatever reason, creates an imbalance of energy, velocities, or revolving mass imbalance, it will generate actions to eliminate such imbalance.

Except for the hanging ring, in all the experiments described previously, the spinning of the bodies was artificially created, but in the test with balls their spinning were generated or at least maintained in a natural way by a vertical stream of air that keep the ball suspended (*floating*) in the air. In the experiment with the balls, it was observed that the orientation of the spinning axis is related to the imbalance (eccentricity); the balls with low eccentricity revolve at high speed about a horizontal axis and those with high eccentricity revolve at low speed about an almost vertical axis. In regards to the vertical spinning axis, the ball's spinning is faster when the heavy semi-sphere of the ball is at the bottom than when it is at the top. Let's explain these facts.

When the ball's eccentricity is low the gravity diameter can easily turn over a horizontal axis promoting the horizontal spinning. On the other hand, when the eccentricity is high the much heavier side of the body creates a strong turning effect towards the vertical orientation, so the ball chasing equilibrium/stability spins about an almost vertical axis while in precession. In this case, if for whatever reason the heavy end is at the top, this end will try to turn towards one side and to the bottom. However, the pushing energy of the stream of air passing around the ball realigns the gravity diameter back to the vertical. This aligning energy is transformed into additional spinning when the heavy end is at the bottom.

The magnitude of the spinning and the orientation of its revolving axis vary in relation to the ball's gravity diameter, implying that there exists a direct relationship between the revolving of a ball and its revolving mass imbalance. In other words, the spinning is induced by the pushing air and the ball's eccentricity. If the *imbalance* of mass and the spinning were independent, there should be observed multiple spinning axis orientations in addition to horizontal and vertical; moreover, the spinning magnitude should be approximately equal for any spinning axis orientation. By analogy to the balls that are suspended in the air by a pushing stream of upward air, it was also possible to conclude that the Sun radiant energy (solar wind) is striking and also pushing the Earth away from the Sun.

Part C Interaction among distanced bodies

Any changes in nature are such as to make the action least.
Pierre-Louis Moreau de Maupertuis
(1698-1759)

"**A**ccording to Einstein, there is no such thing as gravitational force. Rather, the presence of a mass causes a curvature of space-time in the vicinity of the mass" (Serway and Jewett 1141). He suggested that "gravity is not a force like other forces... Bodies like the Earth are not made to move on curve orbits by a force called gravity..." (Hawking, A Brief History of Time 30). Let's find reasons that explain such statements about the gravitational force.

How far is the Earth from the Sun and the Moon?

In this section we are going to get a better appreciation of the relative distance between the Sun, the Earth, and the Moon. The sketches of them available to us and that we normally detail are not in proportion to the actual values that make us to conceive out of proportion some notions associated to them.

Imagine being seated on a park bench resting for a while and you notice that your pet, an eagle, is captivated gazing towards a fence or tree that is about a cricket or baseball field distance (about 128m or 420 ft) from you. You take your cellular phone and start peeping toward such a spot and have the perception of catching a small animal fluttering on the fence. Thus, you decide it is time to do a bit of more exercise and start jogging towards the fence to find out what animal it is. Once there, between two timbers of the fence and about 330 mm (13 inch) from the ground, struggling to escape from a deserted spider net was a brightly colored, orange with black spots, small ladybug of about 3 mm (1/8 inch). Also and just beneath it, near the edge where the field meets the fence, you spot a multicolor marble of about 11 mm (7/16 inch) of diameter. About 31 mm (1 ¼ inch) above and from the marble, aligned with the ladybug, you also watch a red-fire ant of a size of about 2 mm (0.078 -5/64- inch).

Let's re-visualize this scene renaming and introducing the following characters. Instead of a park bench there is an approximately 1.2 m (4 ft) yellow-red balloon that stands for the Sun, the marble represents the Earth, and the ladybug the Moon. See Figure C1. In this scenario, the ant is several times greater than the huge spaceship shown on the Independence Day movie. With exception of the spaceship size, all dimensions and distances are on about the same proportion as the actual relationship among the Sun, the Earth, and the Moon.

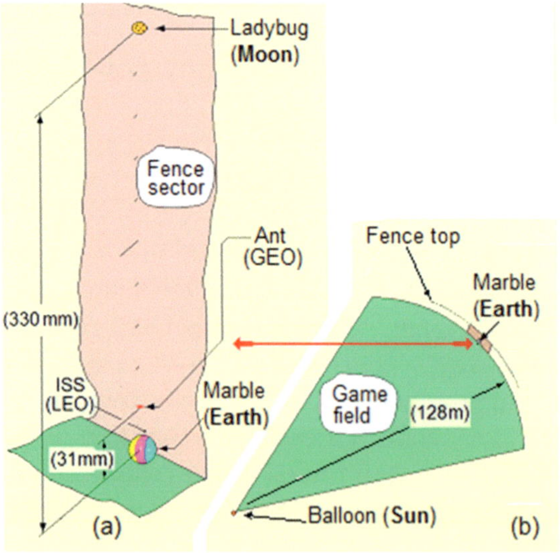

Figure C1. Distance (a) Earth-Moon, (b) Sun-Earth

The distance Sun-Earth reduced gigantically to the size of a game field is a way to see our solar system from one star of a nearby galaxy. To see a game field diminished to a portion of the width of a page of a book, is equivalent to having a view of the field from a commercial airplane that is gliding the city airport waiting for landing authorization. Standing by the fence we can have a naked eyes view, or a close-up of the marble and the ladybug.

The height of the ant from the marble surface is proportional and equivalent to the altitude from Earth to an artificial satellite orbiting at 35,786 Km in the *geostationary earth orbit* (GEO). This altitude, GEO is just about 9.3% of the distance (384,000 km) from Earth to Moon (marble to ladybug). The Hubble space telescope is about 560 km (350 mi) from Earth, or about 1.5% of the GEO altitude. The *International Space Station* (ISS), the largest human-made body, is even closer to Earth, and orbiting between 330 and 435 km (205 and 270 mi), known as the *low earth orbit* (LEO), which is about 1.2% of the GEO altitude, and about 0.011% of the Earth-Moon distance.

The distance between Los Angeles (CA) and Las Vegas (NV) in the USA is about 275 miles (approximately LEO distance). The driving time between these cities is approximate 3:15 hours. As said by Sir Fred Hoyle, British mathematician and astronomer "Space isn't remote at all. It's only an hour's drive away if your car could go straight up."

LEO as part of the outer space is perceived as an altitude far away from us. However, it is so close to us that we, in the book, cannot distinguish it on the scale of the sketch of (*Fence sector*) Figure C1a. In this figure the LEO distance is about 0.06 mm (nil) from the marble surface. This reduced value (0.06) came from the following: depending on the size of the sketch displayed, a distance equivalent to the field length (128 m or 128000 mm) should be about 40 mm[15] on a book page, meaning it has been reduced by a 3200 factor (128.000mm/3200 = 40mm). The marble-ladybug (Earth-Moon) distance also reduced 3200 times is just a tiny fraction of a millimeter (330mm/3200 $\simeq$ 0.1mm), or just a dot in

[15] This 40 mm (representing the 128 m) could become 20 mm (0.78 in) in a screen of a small eReader. The other dimensions shall be reduced in the same ratio also.

the sketch of (*Game field*) Figure C1b. However, on (*Fence sector*) Figure C1a, this dot, or the actual marble-ladybug distance is about 55 mm on the page, that is equivalent to a reduction of six times (330mm/6 $\simeq$ 55 mm). While the marble-ant (31 mm) distance, altitude GEO, reduced six times should also be roughly 5.2 mm (31/6 = 5.2), which 1.2 % (LEO altitude) is about 0.06 mm (5.2x1.2/100 = 0.062). So, according to the proportions of the two sketches of Figure C1, the ISS in the outer space is practically embedded in the Earth.

The bonding between the Sun and the Earth

Suppose that a strong enough magnet is located north facing up at the marble (Earth) position, and assume it is capable of barely attracting the needle of a compass located 330 mm (13 inch) away from it on the ladybug (Moon) position. Again, see Figure C1a. Also, the magnet would strongly attract any small magnetic object at a closer distance, namely at the ant GEO (5.2 mm) distance or at LEO which is at 0.06 mm from it. In other words, at a distance practically inlay in the marble. Similarly, if the Earth's gravitational force attracts the moon, a big object at a far altitude, it should also attract all much smaller terrestrial objects at very close altitude, including, the International Space Station (ISS) and all its occupants. They are practically touching the Earth. Based on the law of gravitational pull, the occupants will always be forced to contact the ISS side facing the Earth due to its attraction. As gravity does on us, it forces us to contact the floor of an airplane when we are traveling at "high" (35,000 ft) altitude and a "high" (560 mph) velocity[16]. However, per NASA pictures and movies, the astronauts float on random directions in the ISS. How is it possible? If Earth's gravity is acting at the short LEO altitude, that should not happen. Does it mean that the Earth's gravity does not exist on the outer space as suggested by Einstein? It seems that it does not exist! Or

---

[16] The cruise velocity of a popular Boeing 747 is about eight times greater than a regular car's.

it is almost nil. The accepted argument of the floatability of a man in the "outer" space said that the velocity of the ISS is so huge that it overcomes the gravitational attraction because everything has to move at the same velocity than the ISS (like when we move in an airplane). This reasoning is comparable to sweeping your hand across a candle flame. In this scenario you do not get burned, but if the flame width is a several miles long (like the vast outer space) the transverse velocity across the flame has to be almost infinite to not get your hand disintegrated. Is the ISS velocity really so huge (close to infinite) to avoid the effect of the gravitational attraction, that is acting with the same intensity in any spot of the outer space at the same height from Earth's surface? If the orbital (tangent to Earth's surface) velocity of the ISS were really so huge, it should take the ISS out of the attraction of Earth like an ant is swept away from any solid surface in the soil for a strong wind. The ISS velocity is much higher than a car velocity, but also it is true that the Earth-ISS distance is (nil) also much smaller (just 0.011%) than the Earth-Moon distance. Therefore, what keeps the Earth and the Moon in orbit? In further sections an explanation is given.

Vacuum, and the Earth-Moon system

An experiment to create vacuum was done using a transparent plastic canister and a balloon whose nozzle-mouth attached to an orifice on the canister lid, was open to the atmosphere. On attachment of Figure C2 is shown an outside view of the connection between the lid and the balloon mouth. For being interconnected through this mouth, the air inside the balloon and the one surrounding the canister are at the same pressure.

What keeps the balloon "inflated" is the sucking of the water and the air in the canister that are under vacuum. For additional details, see Section *Water, air, vacuum, a balloon and a canister*. In this canister, the water and the air are under stretching stress by the internal vacuum. However, each of them remains as a whole by their internal adhesive force(s) that keep them as unique substances. Similarly, every terrestrial thing, object, substance, animal, or living thing, including humans, for sharing this planet, are affined and we all are under the adhesive force, or gravity, that keep us as a whole with our world. The vacuum in the canister is proportional to the resistance of the balloon's expanded membrane. If the membrane's resistance were uniform, the balloon would adopt a spherical shape instead of the irregular heart shape shown in Figure C2. This experiment corroborates that pressure, vacuum and adhesive forces can coexist in a medium.

Figure C2. Water, air, and a balloon in a vacuumed canister

The spherical shape of our planet is developed, I think, by the vacuum of the outer space that permeates through the membrane(s) enveloping the Earth and allows everything on Earth to be under its effect too. We only perceive (positive) pressure due to the agglomeration of air above us. Effect that is similar to diving in a swimming pool: the deeper you descend the higher the pressure in your eardrums.

The pressure, temperature, and density of air decreases as the altitude increases. All available air mixed with water vapor (moisture), dust, salt, and smoke, is produced and kept recycled here close to the Earth's surface. The volume available in our spherical world is much greater at high altitude, where the air quantity is less, and the vacuum is high. The outer space is an almost perfect vacuum at about 3 °K (-454°F), where there is no air. For that reason, the astronauts in the *International Space Station* (ISS) have to produce their own air.

This coexistence of vacuum and pressure promotes the floatability of objects. When pressure is not present, the bodies have the same floatability regardless of their mass, weight, or density. For instance, when a bunch of feathers and a bowling ball under a high vacuum are released simultaneously from the same (about three stories building) height, both reach the floor at the same time, for details see an internet documentary titled *Galileo's Famous Gravity Experiment*, (Cox 18 March 2017). The falling down of these bodies means that the force of attraction of the Earth is still acting over them even on a high vacuum.

The vacuum in the outer space is what allow the planets, stars, galaxies, and other bodies to "float" or remain on relative fixed distance among them. It also accounts for the density (quantity of something per unit volume, unit area, or unit length) of the bodies. On a body of low density there are voids (vacuum) among their molecules. Similarly, there are voids (vacuum) among the floating electrons and nucleus of one atom.

It is known that the Earth is enveloped by strata (known as troposphere, stratosphere, etc.) of different characteristics with the exosphere being the farthest, which starts at about 560 Km (350 mi) above sea level and reaches about 28800 Km (18000 mi). The strata are separated by transition interfaces (tropopause, stratopause, etc.) that are invisible to us, but just as the air, the strata and interfaces are also material things or energy. Proof of this is the heating of the space shuttle during its re-entry to the Earth's atmosphere at an altitude of about 122km (76 mi). To understand how all these earth's atmospheres are very, very close to the Earth, see (Fence Sector) Figure C1a. Furthermore, from the Sun-Earth distance scale, even the perspective of our "far away" Moon is practically touching the Earth. The contiguity principle states that things that occur near each other in time or space are really associated. The Earth-Moon can be considered just one system enveloped by a globe membrane, let's name it the muffle-membrane. See Figure C3.

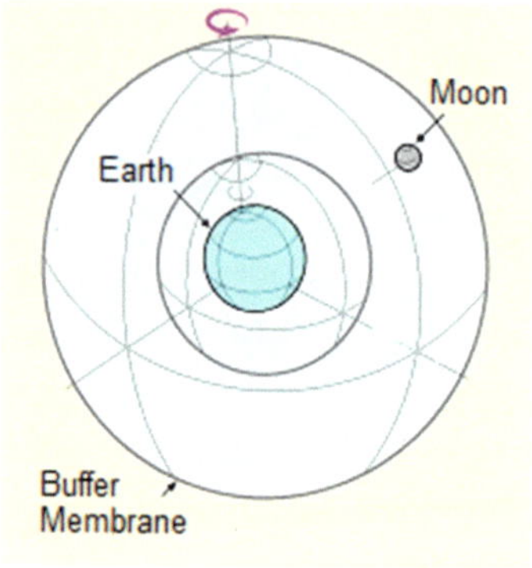

Figure C3. Earth-Moon system

In the Earth-Moon system, it appears that they are on precession about an axis that is quite parallel to the Sun's rays as the air suspended balls do about an almost vertical axis. The night and day cycle is evidence that they are also spinning about and an axis perpendicular to the Sun's ray as the small balls do about a horizontal axis.

It is undeniable that an attraction (perhaps gravity, adhesion, Earth-bond?) keeps us and things against the Earth's surface, but what about in the surroundings near to the Earth known as outer space?

The elliptical path, and forces on Earth's orbit

We are certain and have scientific evidence that our planet orbits the Sun on an elliptical trajectory. To see some traits of this oval path, let's describe one of the simplest, well known methods used to construct an ellipse. The procedure is as follows. Take two nails and nail them separate over a piece of flat wood, and encircle them with a loop of twine, whose length should be greater than twice the distance between the two nails. Inside the loop insert a pen or pencil and dragging the twine with it, move it away from the nails.

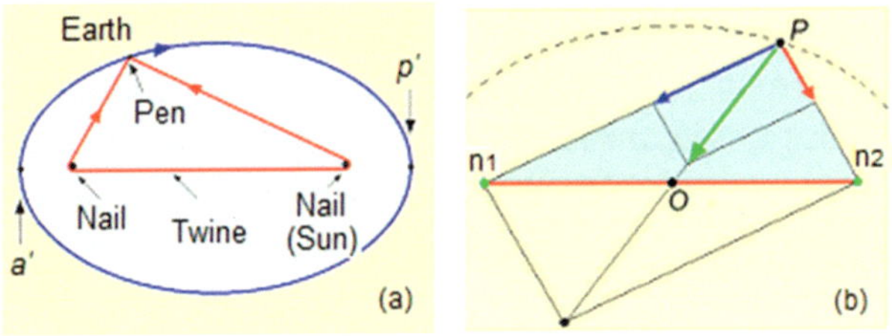

Figure C4. An ellipse curve

Keeping the twine tense, displace the pencil around the nails to draw an ellipse on the wood or on a paper attached to it. See Figure C4a. Keeping the twine tense, implies that whatever the position of the pencil, it is always pulled from two directions aiming toward the two nails. The construction process of the ellipse suggests that it is associated to pairs, one of nails (known in mathematics as foci) and other of tensions or forces. The pencil (P) and the nails (n_1 and n_2) form a triangle that is a half of the parallelogram whose one of its diagonals is the distance between the nails. See Figure C4b. Knowing that *the diagonals of a parallelogram bisect each other* (see Figure A1d), through the nails' midpoint (O) shall pass also the other diagonal coming from the pencil. Assuming the forces from pencil towards the nails are proportional to nails-pencil distances, the resultant of such forces shall also aim to the nails' midpoint.

A well-known case of a body describing a curved path is a flying projectile, whose trajectory describes a parabola curve because its horizontal velocity, usually, is constant while the vertical one is different and variable for the attraction of Earth. Other example of an object moving in a curved path by duality of actions (forces) is the skipping bomb diving while spinning in the water. In these two examples, the direction of the two actions promoting the curve, are perpendicular. The curved path should be something other than a circle if the magnitude of the two velocities were different as happens with a projectile in motion.

This requirement of duality of forces is similar to whirl in the air a stone tied at one end of a twine. To keep it moving in circles[17] and at uniform speed, two movements are indispensable. One is a lash away from us as if we were using a whip, and the other a small whirl impulse timed or synchronized just ahead of the stone. That we usually apply to it just when it is in front of us. Also two forces keep it moving in a closed curve, namely the centrifugal force created by the whirling motion aimed away from and counteracted by the holding hand through the twine, and a tangential force created by the angular (impulse) momentum. However, we have been told that while the Earth is orbiting the Sun, it is attracted by only one gravitational force!!! It seems that something is missing in this conceptualization. What makes the Earth move around? It has to be a lateral force perpendicular to another force acting in direction of the straight line between the Earth and the Sun.

The gravitational force might make the Earth rotate because it creates a torque on it acting in its eccentric center of mass. But it does not make it orbit because it is perpendicular to this trajectory. Why is Earth rotating and orbiting? What makes the Earth do these two movements?

Figure C5a is a close-up of the pulley of the pail hanging on a bracket shown in Figure A2. When the pulley and the ropes were set and the pail was not suspended yet, the center (B') of the pulley was vertically aligned with its hanging (A) point. See Figure C5b. Once the pail was attached to the rope and set free, at that instance an imbalance was generated and the pulley started moving to the left searching equilibrium. Once it was reached, the

---

[17] When doing this experiment, observe that your fingers or hand that hold the twine should be moving in small oval trajectory. If they were fixed on a point, the curve would certainly be a circle but you cannot accelerate the stone.

movement ceased. This is another proof that a lateral force away from the straight line between the Earth and the Sun is also needed to create the lateral or orbital movement of our planet around the Sun.

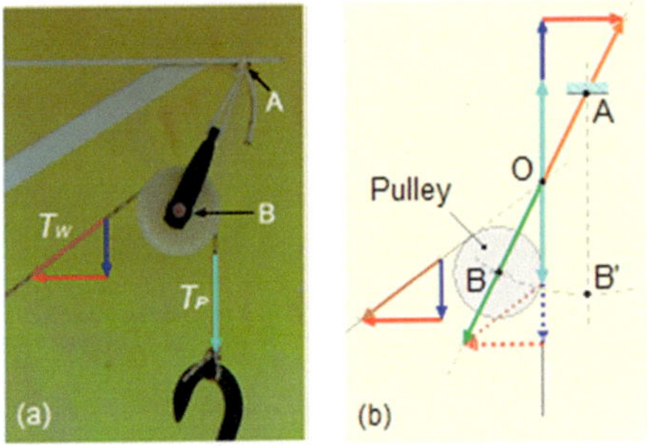

Figure C5. Forces on a pulley

The axle or (B) center of the pulley shown in Figure C5a is not aligned vertically to its hanging (A) point because there is a horizontal and lateral force that pulled the rope and moved the pulley to the left against the wall. Like a rope in a tug-of-war, two tensions of equal magnitude and different directions are exerted on the ends of the bottom rope partially wrapping the pulley, one (Tw) by the wall and the other (TP) by the hanging pail. When the direction lines of these tensions are prolonged, they meet on one encounter (O) point that is over the top rope. Figure C5b. The projection of all the forces acting on the static pulley, including the results of the tensions of the ropes, converge into this point.

The pulley is in equilibrium and it is not moving because the top rope hanging the pulley counteracts all the vertical and horizontal forces [18] generated by the bottom rope. See the similarity between this pulley whose rope tensions are balanced in this encounter (O) point, and the paint can handle that is balanced in its center of mass as it is shown in Figure A4c. These two objects

---

[18] The pulley and ropes weight are not included. They are nil compare to the pail weight.

are in (static) equilibrium because the forces acting in such points are balanced. Similarly, if all the external forces acting on the Earth-Moon system were concentrated or converging to its barycenter (point equivalent to O point and center of mass, see Part E), this system should be static regarding the Sun or other far star as the pail is on regard to the ground.

Like the vertical and balanced forces acting over the pulley, in order to be approaching and getting away from the Sun, two opposite and balanced forces should be acting on the Earth and over the straight line between the Earth and the Sun. These forces should be acting in the Earth-Moon barycenter like the tensions acting in the point O of the pulley-pail system. If the gravitational force attracting the Earth towards the Sun were one of these forces, where is the other one counteracting it? On *Effects of energies acting on the Earth* section this will be revealed.

The driving and striking energy from the Sun

The warming energy or radiation received by a person from a bonfire depends on how hot, vivid, and big the fire is. The amount of the energy taken, depends on how close (more warming) or far (less warming) the person is from the flame.

This energy is inversely proportional[19] to the square of the distance to the fire. This relationship is known as the inverse square law. In other words, if you are comfortable at one meter from a bonfire and it radiates to you 9,000 units of warming[20]. When the distance is three meters, you may be at the mercy of the cold weather because you will receive a ninth (one divided by three to the second power) of it or 1,000 units of heat. On the other hand, if you instead take a step toward the flame and reduce the

---

[19] Two magnitudes are inversely proportional when one of them increases as the other decreases, and they are linked on a directly proportional way when one becomes greater the other too.

[20] Typical units of heat radiation rate or power are Kw (kilowatts), J/s, and Btu/h.

original distance to just two thirds of a meter, it will warm you with 20,250 units, and it could be unbearable for you.

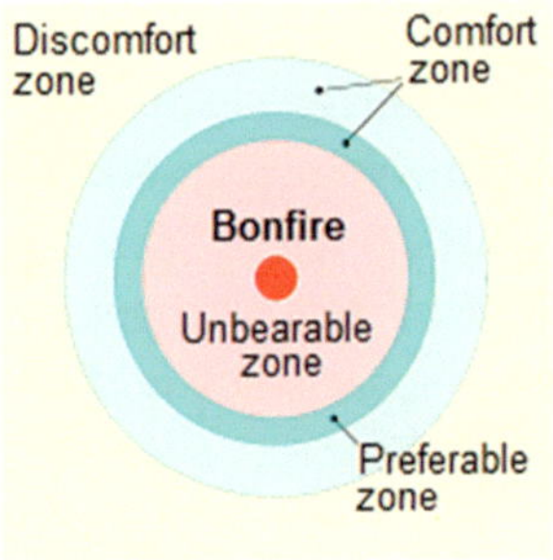

Figure C6. Bonfire radiation zones

Figure C6 shows the typical zones around a bonfire. The relative wide comfort zone has a warming, preferable, and narrow band near to the fire which internal edge is very close to the threshold limit with the unbearable zone. Two or three steps from this threshold towards the fire and in a few seconds we need to go back.

Based on Stefan's law (an inverse square law), the scientists have grouped those radiation parameters (distance from the flame, its temperature, etc.) between two sources or bodies (bonfire and you, or hot and "cold" bodies) in one formula (Kern 79), which can be simplified as

$$E_r = K\,(T_h^4 - T_c^4)/r^2$$

Where K is a constant. If the Sun and Earth absolute temperatures are pretty fixed, the term or the difference in the parenthesis remains also fixed and can be included into the constant. In addition to the new constant, what is left is the distance (*r*) between the bodies, and the net radiant and bidirectional energy (E_r) interchanged by the bodies or how much energy emitted by one of them reaches the other, and vice versa. Comparing the Sun and Earth is equivalent to comparing a big apple (Sun) against the writing ball of a ballpoint pen, and they being separated a distance of about 10.6 m (11.6 yard). Hence, in this couple, it is safe to say that this interchange is just the energy received by the Earth because none of it returns to the Sun. Therefore, what the above equation said is that the energy received (E_r) by the Earth is inversely proportional to the square of the distance (r^2) to the Sun. As the Sun-Earth distance is not unique and it describes an elliptical shape, the energy received by the Earth should be elliptical also. However, they are inversely proportional, so when the distance is decreasing the energy is increasing, and vice versa.

We are certain that during a year the Earth is closest[21] to the Sun at the beginning of January and farthest from it at the beginning of July, consequently there is an Earth-Sun distance which should be close to the average. For explanation purposes, it will be called median distance, through which the Earth receives enough energy to keep it orbiting, spinning, and life is safe in its surface. If a distance is shorter than *perihelion* (known as the recorded nearest distances from the Earth to the Sun) or greater than *aphelion* (known as the farthest distance from the Sun), life would not be possible.

Assuming it has reached its median distance, and according to the equation presented previously, a small reduction of the Earth-Sun distance will increase on a big amount the energy received from the Sun. Part of it will be absorbed by the Earth itself, and the rest should be used to transform the Earth's momenta. Will it be the linear or the angular momentum, or both? This question is answered in *The Sun striking energy and the Earth's orbit* section.

Effects of energies acting on the Earth

The received energy from the Sun is continuously absorbed by the photosynthesis of plants, warming of oceans and land, etc.

---

[21] Seasons are due to the tilt degrees of our planet spin axis. For that reason seasons are reversed in the hemispheres.

It is also dissipated by the cooling nights, breezes, moving of the water of the oceans, among other mechanisms. For the irregularity and complexity of our planet (mountains, jungles, deserts, lakes, cities, etc.), the distribution, transfer, and absorption of energy is not perfect, and some of it remains as an excess (ΔE) on the receiving side, creating an energetic imbalance with the other shade-side of the Earth.

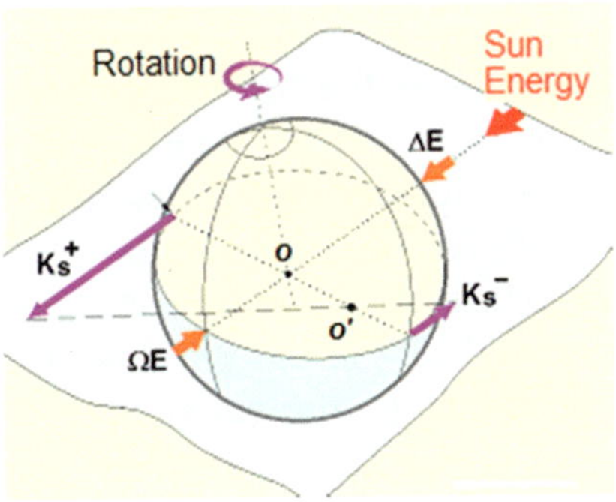

Figure C7. Dynamic imbalance in Earth

The received energy and its remaining excess in the Earth are pushing and oriented on the same direction as "the solar wind[22] that makes that the tail of a comet always points away from the Sun, whether the comet is headed towards or away from the Sun" (Wikipedia, Solar Wind - History, 4th paragraph). We observed a similar pushing effect on the experiments with balls detailed in *Part B Rotating bodies chasing dynamic equilibrium*.

To get balance, nature or the energy itself creates a compensating or reacting energy (ΩE) acting towards the Sun or on the opposite direction to the coming one. See Figure C7. We know the effect of this compensating energy as the gravitational force. The existence of this kind of pushing energy can be observed

---

22 A stream from the Sun charged with particles like electrons, protons, and alpha particles, radiated as heat and/or light rays.

in action by spinning a toy gyroscope as it is detailed in *The spinning toy gyroscope* section.

In the Earth, these energetic imbalances become evident in several ways, one for instance is during its rotation. Viewed from the Sun, an eastern spot (Ks^-) of the Earth in the last twelve hours on the nightshade is relatively cooler in the morning, than a western spot (Ks^+) that has been on the sunlight the same twelve hours. This difference of energy creates an instantaneous center of gravity, or dot (o'). See Figure C7.

The center of mass of the Earth-Moon system is located at about three fourths of the Earth's radius. The location of this center creates an eccentricity that is variable by the constant approaching and getting away of the Moon to the Earth. This is another source of imbalance. However, based on the conclusions devised with the experiment of the split key ring spinning about one of its vertical diameters, the writer believes that nature is searching a dynamic equilibrium during these Moon movements. This topic is detailed in *Moon's role on Earth's equilibrium* section.

The imbalance of energies in the Earth and its rotation promote the orbiting movements of our planet as the energetic imbalances and the spinning do on the toy gyroscope and on the offset-hanging spinning wheel. These factors create a force that makes them orbit in search of equilibrium. In such experiments, the lineal movement occurs on a direction that is opposed to the one that the body (wheel) would move towards if it were rolling over a surface. That is, from the geometric center or *static* center of mass towards the instantaneous center of mass. The magnitude of the created (lateral) force is proportional to the magnitude of the imbalance promoting such movement. For details about this lateral force see *The spinning toy gyroscope* and *The hanging spinning wheel* sections.

The unbalanced Earth including its buffer-membrane searching for more stable conditions, keeps chasing the instantaneous center, point o' in Figure C7. This action generates, and promotes the Earth's lateral or orbital movement which is perpendicular to the energized rays coming from the Sun. Simultaneously, as a consequence of the reacting energy (ΩE) that compensates the excess not absorbed of the Sun's striking energy, the Earth is also moving continuously towards the Sun. As this centripetal velocity is always aimed towards the Sun, we have named it the *sun-aim velocity*. The next section enumerates the relationship between these velocities and the striking energy received from the Sun.

The Sun striking energy and the Earth's orbit

Similar to a ping-pong ball on an upward air stream, the Earth-Moon system, and probably its buffer-membrane also, started to rotate many years ago due to the *pushing or striking* energy coming from the Sun. In addition, this rotation was and is promoted by the changeable Earth's eccentricity, and by the Earth's own angular momentum.

The momentum of a body, such as the Earth, is defined as the product of its mass and its *distinctive* (angular or linear) velocity. The linear momentum of the Earth is its mass times its linear velocity, while its angular momentum is equal to its mass times its angular velocity times its moment of inertia that is a positive big factor. As the mass of the Earth is unique, and the only other variables in its momentum equations are its angular and orbital velocities, we can just talk about these velocities to make reference to the corresponding momenta. To compare these unlike velocities[23], and based on some conservative suppositions, were found three ratios that are far from accurate but that give an educated estimate of the order of magnitude of these two quite different velocities. For details see Section *Approximate calculations associated to the Earth*. According to the most conservative ratio, the linear speed of the Earth is about 64.3 times greater than its *transformed* rotational speed. The Earth's orbital velocity is about 107302 km/h while its rotational velocity (giving

[23] Actually it is not possible to compare these two dissimilar velocities.

a revolution on a day), is about $0.2618\, Rad/h$. These values and the equations of momentum reported above were used for double checking magnitudes. It is found that the angular momentum of the Earth is much greater than its linear (orbital) momentum.

Both, the experiments with the spinning wheel and the toy gyroscope, and the assumption made about the transformation of rotational speed into linear speed, seem to be in agreement that the angular momentum of the Earth (around its spinning axis) is very much stronger than its linear (orbital) momentum around the Sun. The linear momentum is the weaker and as such is the most susceptible to vary. So, the orbital or lateral speed is the first and probably the only that varies. If the sun-aim velocity were the one that varied, the analysis would be quite similar to the one detailed in further paragraphs, however if the two velocities vary simultaneously, the analysis would be much more complicated. The variation will occur gradually because the massive Earth does not allow sudden changes.

Due to the gigantic distance between the Sun and the Earth, the orbit of the Earth is almost a perfect circle, just a bit oval. The orbital velocity of the Earth is tangent to its orbit and parallel to the always tangent lateral velocity. Actually, these velocities become just one. For that reason, the sun-aim velocity and the lateral (orbital) velocity are perpendicular [24] between them. However, on point *B* of the sketch of Figure C8, they appear not to be, because in the sketch the oval shapes and the almost equal magnitude of the velocities were exaggerated for explanation purposes. This figure combines into one sketch the received energy and Sun-Earth distance. They are read as follow: for instance, when the Earth is at distance *A*, the energy is *m*; at point *P*, energy *M*. In general, when it is at any point of its orbit at a distance *r*, its energy is *Er*.

---

[24] Two vectors, lines, or planes are perpendicular if they meet not obliquely with equal space (angles) on either side. On a building the (verticals) walls are perpendicular to the (horizontal) floor.

In future paragraphs, the velocities and their magnitudes are worded as just lateral or sun-aim velocity. Similarly, when the Earth is at one point or position, for instance at *A*, its distance to the Sun is worded either as distance *A* or just *A*, and the received energy at whatever location (*r*) is just energy at such point.

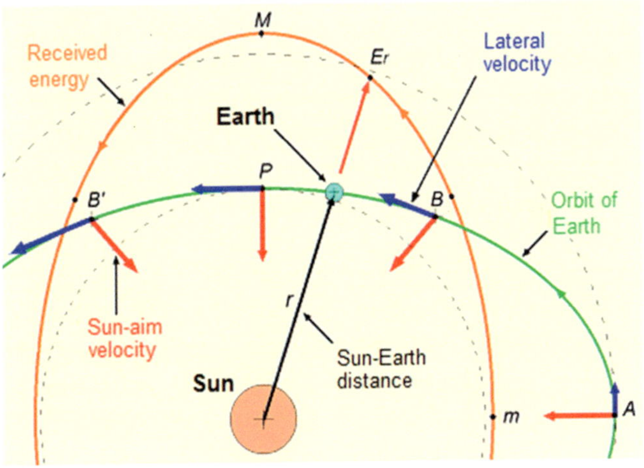

Figure C8. Earth-Sun interactions

In the classical conceptualization of the elliptical Earth's orbit in which the Sun is at one of the ellipse's foci (right nail in Figure C4 or about *m* point in Figure C8), the closest and the farthest Earth-Sun (perihelion and aphelion) distances are defined when the Earth, Sun, and the ellipse's two foci are aligned. It means that the Earth is only one time at each of such distances. However, in this text, the Sun is supposed to be located in the middle of the two ellipse's foci, as it shown in Figure C8. So, the Earth is twice at a very far (*A*) distance from the Sun, and twice at a very close (*P*) distance. The (A's) points of one set are aligned with the foci, and the others (P's) are not. Only one point of each of them are shown in the sketch. This duality of A and P points explains why the four stations occur twice in a year at different times and places of our planet.

The Earth will *approach* or move towards the Sun, when the magnitude of the sun-aim velocity is greater than the magnitude of the lateral velocity, and when this difference is on the other way around, it will *drift apart* or move away from the Sun. The magnitude of the lateral velocity gets its two lowest values when the Earth is at any of the farthest distance ("aphelion") from the Sun, or when it reaches an *A* points. See Figure C8. At these distances the received energy is at its minimum (*m*). When the Earth is at any of the closest distances ("perihelion") or any of the two *P* points, these two velocities have the same magnitude, and the energy reaches its maximum value (*M*). In other words, the orbital velocity is inversely proportional to the striking energy received from the Sun, and the received energy is inversely proportional to the Sun-Earth distance.

It is worth to note that during a complete orbit, the Earth goes twice through two different, consecutive, and alternative stages that are part of half the complete orbit trajectory, namely *approaching* the Sun that starts at "aphelion" (*A*) and finishes at "perihelion" (*P*), and then *drifting apart* from perihelion to the next aphelion. These two stages are also present on the received energy curve but the reaching of the maximum and minimum energy occur in opposite directions as it is suggested in Figure C8. To visualize and describe these stages, imagine that the Sun is a bonfire whose comfort zone is limited by aphelion and perihelion or the two circles on the dashed line. Just after an A point, the Earth approaching is at its fastest rate because the sun-aim velocity is much greater than the lateral velocity. However, close to this location there is few (m)[25] energy to be transformed into velocity, so the velocity growth's is slow. Once the Earth is on a location close to *P*, for example at *B*, the approaching is slow, but now the energy, which is still growing at a fast rate, has reached a big value that increases substantially the velocity. When the Earth is at a distance (*r*), between *B* or *B'* and *P*, its level of energy (*Er*) is very

---

[25] Reminder: The farthest from the flame (Sun) the less (energy) warming.

close to the maximum value *M*. So, from point *B* to point *B'*, even though the approaching and the drifting apart are very slow, the growing of the lateral velocity is at a huge rate for the high value of energy. At point *P*, this velocity reaches the same magnitude as the Sun-aim's, and overpasses it by a big amount at point *B'* for the still available amount of energy. Notice that at point *B* the Sun-aim velocity is greater than the lateral's, while on point *B'* is the other way around. On the remaining section of the drifting apart stage (from *B'* to the next *A*), the lateral velocity is drastically reduced to the same magnitude as the previous *A* because both the drifting and the energy are decreasing at high rates. The energy growth is flat or almost constant only when the Earth is at or very close to (*P*), or when the approaching or drifting apart is almost nil. These stages are repeated on the other half of the trajectory, or from left to right on the bottom half not shown in Figure C8.

Any imbalance of energy on the Earth promotes its rotational movement and consequently its angular momentum. Simultaneously, the angular momentum and the imbalance of energy chasing for equilibrium generate the force that induces the Earth's orbital movement and its (lineal) momentum. As the angular momentum of the Earth is much stronger than its lineal momentum, any change on the excess of energy received from the Sun will affect and change the weakest lineal momentum that is conveyed as lineal (orbital or lateral) velocity.

Moon's role on Earth's equilibrium

The Earth-Moon system also uses another mechanism to chase equilibrium. It adjusts its angular momentum by varying its moment of inertia when the Moon changes its distance to the Earth. That the Earth-Moon system does, the writer speculates, to keep its angular momentum pretty constant.

This process of the Earth of getting closer and farther from the Sun by varying its lateral velocity to make it faster or slower than its sun-aim velocity, is like the process of the Moon getting closer and farther from Earth by varying its orbital and rotational speed. "When the Moon is at its perigee, its rotation is slower than its orbital motion... Conversely, when the Moon reaches its apogee, its rotation is faster than its orbital motion" (Wikipedia, Orbit of The Moon - Libration, 1st paragraph).

The Moon is the key balancing media that nature uses to keep the Earth-Moon system on a stable condition by carrying on, promoting, and regulating important counterbalance effects. These effects are similar, although much more complicated, to the movements of the split key ring's mouth that change positions to get a more stable or balanced condition in response to the variations of energy caused by the changes of the RPM of the ring (See Section *The spinning split key ring).* In this system, nature has anticipated all possible unbalancing events. That seems to be periodic and need to be tweaked because they vary their magnitude and/or occur at different conditions and moments. For instance, we have an estimate of the season of hurricanes, but it is almost impossible for us to know accurately where and when they will start and finish, and what would be their exact path and their magnitude.

As most of these disturbing events are periodic on a yearly basis, the orbit of the Moon is also periodic but it is self-adjusted accordingly. For instance, in the year 2014 the orbit (time) frequency was almost exactly between each complete Moon's orbit. However the amplitude among the orbits varied reaching four peak differences between the *apogees* (farthest distance of the Moon from Earth) and the *perigees* (nearest distance of the Moon from Earth).

From Figure C9 (Wikipedia, Orbit of The Moon - Lunar periods, Graph), the two biggest apogee-perigee differences happened at the end of April and at the end of November of that year (about 407,000 km minus about 357,000 km), and the two smallest occurred at the beginning of January and August (405,000 minus 370,000 km). These trends of apogees and perigees of periodic differences are kept for years with small variations, but nature not only tweaks these differences, but also other parameters, like the angle (α) between the Moon's orbit plane and Earth's orbit plane. See Figure C10b. For example, this angle varied from about 5.0 to 5.3 degrees from 2008 to 2010, (Espenak 12 January 2012). The tilted angle (β) of the Moon spinning axis also varies from about 22.1 to 25.4 degrees during cycles of about 40,000 years.

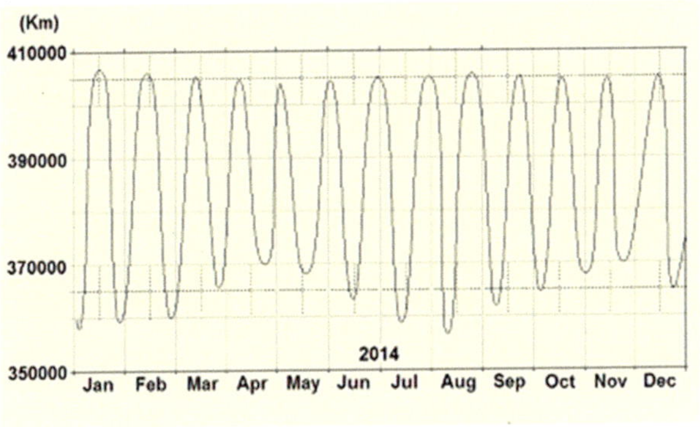

Figure C9. Earth-Moon distances

The Earth spins, wobbles, weaves, and rocks its spinning axis on different directions and positions to expose different parts of its surface to absorb on the most uniform and efficient way, the energy coming from the Sun. Additionally, due to its variable geography and separation, the Earth-Moon system lost its dynamic balance that is recovered mainly by the relocation of the Moon orbiting. However, the Earth system does not remain orbiting over just one plane. Instead, chasing a stable condition, the Moon is switching positions above and under the ecliptic plane.

The apsis line seems to remain parallel during a year in Figure C10a, but actually it is on a gradual counterclockwise rotation (known as apsidal precession) that takes 8.85 years to complete a full revolution. The nodal line, as shown in Figure C10b, is on a clockwise precession having a period of 18.6 years. This movement is named as a nodal precession in reference to the two nodes or points (d and u) where the orbit of the Moon crosses the ecliptic plane.

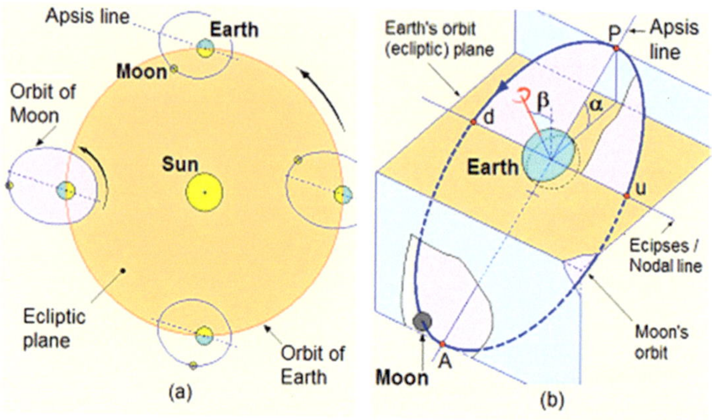

Figure C10. Earth and Moon (a) orbits, (b) Moon's orbit detail

To visualize these two precessional movements, imagine the plane of the orbit of the Moon as an airplane, in a curved trajectory like a vulture gliding over the carrion, approaching the flight deck of an aircraft carrier ship that is also turning or changing its direction, but on an opposite direction to the airplane direction. The movement done to align the airplane with the ship runway is equivalent to the apsidal precession; while the movement to place the winds of the airplane parallel to the flight deck, is equivalent to the nodal precession.

Nowadays the Earth's spinning axis tilts 23.4 degrees to the plane described by the orbit of the Earth around the Sun. It is tilted towards the Sun some months of the year and away from other months. See Figure C10a. As the energy always moves toward a thermodynamic equilibrium, this inclination should be the most efficient arrangement to uniformly distribute all over the Earth's surface the energy received from the Sun. Let's consider three other possible distribution arrangements observed during the experiments with the floating balls, namely the spinning axis perpendicular to the Sun's rays. In this case, the equator and all its near spots will uniformly distribute the received energy. However, the frozen area of the poles would be greater for receiving just slanted rays all year long. If the axis were parallel to the Sun's rays, it would be the worst case scenario because one hemisphere of the Earth would be frozen perennially, creating a big energetic difference between both hemispheres. The worst disadvantage of these two alternatives is that they do not promote the orbital movement. The third option, a tilting of 45 degrees, seems more efficient but it would require a big amount of the unstable wobbling or precessional movement to distribute the energy between the Earth's hemispheres. By orbiting the Earth and by fixing its spinning axis at about 23.4 that happens to be about half of 45 degrees (45/2 = 22.5), nature minimizes the precession.

A summary about Sun, Earth, and the Moon.

From the similitude with the emission of heat from a bonfire, and the intensity of illumination radiated from a bulb, it was deduced that the energy emitted from the Sun and received by the Earth follows an inverse square law. It states that the intensity of the radiant energy on a receptor (Earth) is inversely proportional to the square of its distance from its radiation's source (Sun).

Our planet is not symmetrical even when it appears and looks spherical when it is seen from the outer space, and its mass is not uniform due to the complexity of its geography and variety of elements that made it. The Earth and Moon due to their separation, as a system is eccentric and definitely is neither uniform nor symmetrical also. These differences not only impede the center of mass of the Earth from coinciding with its geometric center, but they also avoid that it gets to a steady equilibrium when it is moving, especially when it is spinning.

These imbalances (energy and revolving mass imbalance) create at least two forces on the Earth, one on the orbital direction, and the other, centripetal force known as gravitational force. This is directed towards the Sun to counterbalance the centrifugal one, away from the Sun, generated mainly by the striking and pushing effect of the Sun's radiant energy (solar wind). From the experiments with suspended balls, we know that the pushing up of the air not only balances the gravity but it also promotes the spinning of the balls as a way to chase equilibrium. Similarly, the striking and pushing of the radiation from the Sun promotes the spinning of the eccentric Earth. We are certain the Earth is moving in orbit, this velocity should be generated by the orbital force as it was corroborated in the experiments with the toy gyroscope and the bicycle wheel.

A curved trajectory, like the ones of a projectile and the diving cylinder, is formed by a velocity tangent to the body's path, and a radial velocity aiming to the center(s) of the curve. These two velocities, the energy received from the Sun, and the Earth's imbalances modulate the Earth-Sun distance. The difference (magnitudes and directions) between these velocities generates the elliptical trajectory that promotes and varies the quantity of solar energy received by the Earth. The remaining or the excess of energy varies continuously because the constant changing of the Sun-Earth distance and Earth-Moon distance, so this imbalance never reaches an equilibrium making the chasing process perennial. It is this changing imbalance that keeps the Earth rotating and orbiting, chasing the never reachable dynamic equilibrium.

What really keeps the planets orbiting around the Sun, their main source of energy, and electrons orbiting around its nucleus is just an eccentric distribution of their mass combined with an imbalance of energy (mainly their spinning and/or lineal or orbital momentum) in a constant chasing for a steady dynamic equilibrium.

In order to maximize the received energy and minimize its excess, two main factors are involved: the spinning of the small body about an axis passing through itself, and the orbiting of it around the big source of energy. Both movements promote a uniform interchange of energy. When a small spinning body orbits around a big object, and its spin axis is in precession it exposes most of its surface to the other object, and while in orbit, it gets exposed to the maximum possible surface of the big one.

These phenomena (orbiting and spinning) and everything associated to the transmitter and receptor of energy obey a universal law, which is the cycling law of energy. We are aware of it on the atoms, the solar system, weather stations, days cycling, and on our working day routines, etc. In general, it seems that everything around us at all levels is a perpetual cycling whose period[26] may vary, depending on the size of the object, from a fraction of a second to millions of years. Cycling and circles are embedded on us, and on moving (rotating-spinning) or static things around us, as we observed when we bend a flexible solid strip like the one of the *A flexible ruler and an invisible wheel* experiment.

[26] Period is the portion of time between a complete revolution of a recurrent event.

Part D Homemade experiments in details

The most exciting phrase to hear in science, the one that heralds new discoveries, is not "Eureka" (I found it), but "that's funny..."

Isaac Asimov

Galileo and Newton "believed that a few key or critical experiments would yield correct fundamental principles .... The way to obtain correct and basic principles is to pay attention to what nature says rather than what the mind prefers... Galileo decided that one must penetrate to what is fundamental in phenomena and start there... He had observed that the speeds with which dissimilar objects fall differ less in air than in water. Hence the thinner the medium, the less difference in speed of fall among bodies. Having observed this I (He) came to the conclusion that in a medium totally devoid of resistance all bodies would fall with the same speed." (Kline 330-332).

The sections of this part are mainly related to simple and home-made experiments (tests or trials to demonstrate or confirm facts, most of them known) that can be made using common and handy things.

For the future sections you must have a basic understanding of algebra. Actually you might not even need to know that. In algebra we can talk about numbers or things we do not know by assigning them short names known as variables like *a* for apple, *d* for distance, *x* for position from vertical axis, *y* for position from horizontal axis, and so on. The equals sign means that two things that appear different are the same. For instance, $I = M.\ r^2$.

A Split key ring spinning about a vertical axis

If a coin is tossed 100 times in the air, or it is spun on a table, the chances of getting it face-up or face-down is the same, meaning that about 50 times it will be face-up and 50 times face-down. A small difference between the two cases is considered acceptable. What effect would there be on the probability of getting the coin face-up if one of its side were just a bit heavier than the other? This question inspired the experiment with split key rings spinning on a flat surface.

Two identical composed rings were prepared for this experiment, one with a "big" bead lodged on a fabric silky thread, and the other with a small bead mating with a human hair. Table D1 summarizes the data of the main components of such rings. Figure D1 summarizes all the types of rotational movements of the split key ring observed during a stable rotating process.

Table D1. Data of components of composed rings

COMPONENT	WEIGHT (gr)	DIAMETER (mm)
Split key ring	6.05	39
Big bead	0.033	3.9
Small bead	0.0068	1.1

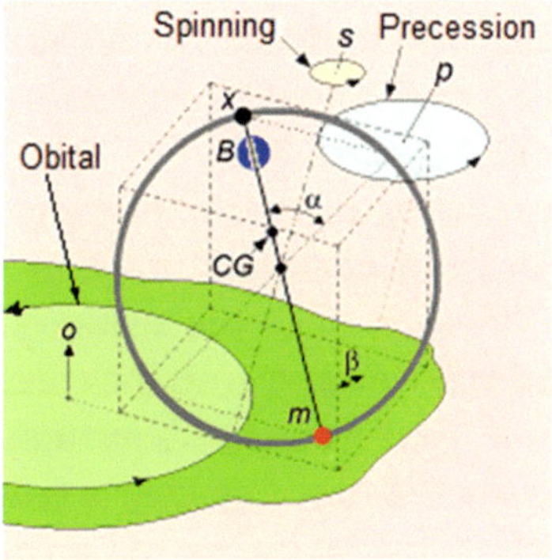

Figure D1. Split key ring motions

The split key ring has a uniform weight by unit length except by its mouth (section through which the keys are placed in the ring) that is formed by only one ring while the rest of the split key ring, is formed by two. So, the mouth spot is lighter and weighs half grams per millimeter less than any other ring zone with the same length. Due to this weight difference, the center of gravity (CG) of the split key ring is away from both its geometric center and the mouth. The distance between these two points is named eccentricity in this book. However, when a free-sliding bead is placed on a string coinciding with the mouth diameter, the center of gravity will shift its location along the thread depending on the bead's weight and location.

The mouth *m* of the ring is also identified with a red dot while its contra-mouth (*x*), is identified with a black dot in Figure D1. These spots are part of the diameter (*mx)* that is unique for having ends with different weights, and for splitting the ring weight into two equal parts. The ring's center of gravity (*CG*), even when it is not visible, is always located in this called gravity diameter. In this diameter are aligned the mouth, contra-mouth, the ring geometric center, and the center of gravity (for clarification about this fact see *Balancing and the add-cut theorem*). The straight line connecting this diameter represents the string; the dashed line, the spinning *S* axis. This axis, could be a bit slanted or inclined a small angle "β" from the vertical. The small (dark blue) globe stands for the bead *B* that is free to slide along the string, which forms an angle α with its mirror one.

Most of the spinning were initiated with the string on the vertical position and the contra-mouth and the bead on the bottom position. However, once the spinning started, the bead could be either next to the mouth or to the contra-mouth, or in an intermediate position. The Figure D2 are pictures of a spinning split key ring having the small lead sliding over a taut string connected to one of its diameters limited by the red painted mouth and its diametrical opposite point or black-painted contra-mouth spot.

During the gradual decreasing of the spinning, the ring passes through a process of six (6) typical stages, five of which are shown in Figure D2. The *third stage*, not shown, is the falling and almost instantaneously sliding down of the bead. It was not possible to take a picture of so short an action. These stages were defined, after hundreds of spinning, according to the position of the gravity diameter. Even though the string and the "big" bead help to identify the mouth and the contra-mouth, it is hard to see and/or differentiate their positions with just naked eyes because the ring can spin above the 1000 rpm[27]. For that reason if you try this experiment, it is suggested to clearly identify the lighter spot (mouth) and the gravity diameter of the split key ring. Then film the spinning, to later watch it in slow motion.

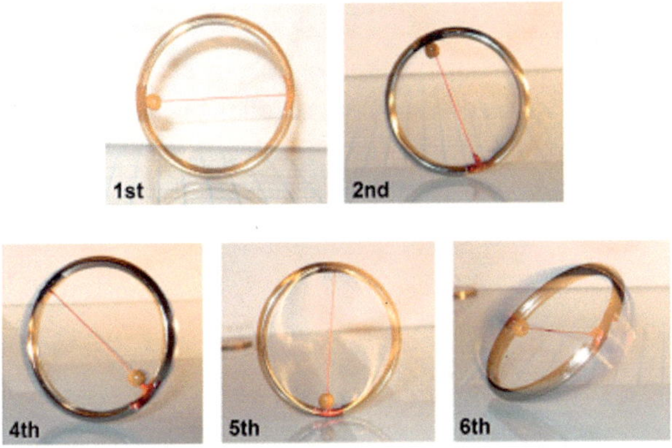

Figure D2. Main stages of the Split key ring rotating process

On the *first stage*, just after the start of rotation, the mouth diameter reaches a horizontal or almost horizontal position that can be distinguished by the thread-bead position, and/or by the red flash color of the mouth. Except on the first and last stages, the heavy contra-mouth reaches the top section of the ring! On the *second stage* the gravity diameter is much closer to the spinning axis, while on the fourth stage is a bit farther without reaching close to the horizontal diameter. The *fifth stage* happens when the fourth stage is not stable anymore. In this stage, the orbital

[27] We cannot see the nuts of a car wheel spinning about 800 rpm.

movement is practically null. There is a special stage, seen on few occasions, when the second, fourth, and the fifth stages become just one, and the string coincides with the spinning axis at the almost true vertical without the precessional and orbiting movements. On the *sixth stage*, the ring begins to stop its rotation and/or starts falling and getting up as if it were rebounding. At the beginning of this stage, it is possible that the tread or the mouth diameter reaches a true vertical position two or three times more.

Why does the contra-mouth, the heavy spot, of a rotating split key ring move up and stays in this top position for a while during the second stage? That is opposite to where a heavy spot of a ball goes when it is placed on a flat, smooth horizontal surface. It is also opposite to what the heavy bulged sector of a bent shaft does when it is placed over two parallel big and straight shafts leveled horizontally.

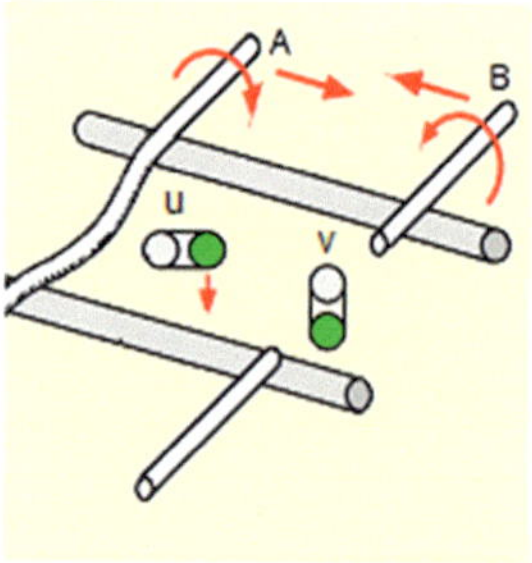

Figure D3. An unbalanced slender long shaft

This is done on an industrial process, known as static balancing, of straightening slender round shafts. See Figure D3. Once placed on the big shafts, and due to its bulge, exaggerated for explanation purposes, a bent shaft starts rolling and swinging back and forth between two points (*A* and *B*) until it reaches equilibrium and the bulge is at the bottom section of the shaft. See detail *v*. Then, the technician identifies and marks this sector, where later he/she applies a force to eliminate or minimize the bend. After that, the process is re-started and repeated until the shaft has been straightened and stops swinging between two points.

When a tiny weight is added to or taken away from the surface of a shaft, its center of gravity does not coincide with the geometry center of its circular cross section anymore. This cross section would resemble the split key ring shown in Figure D1.

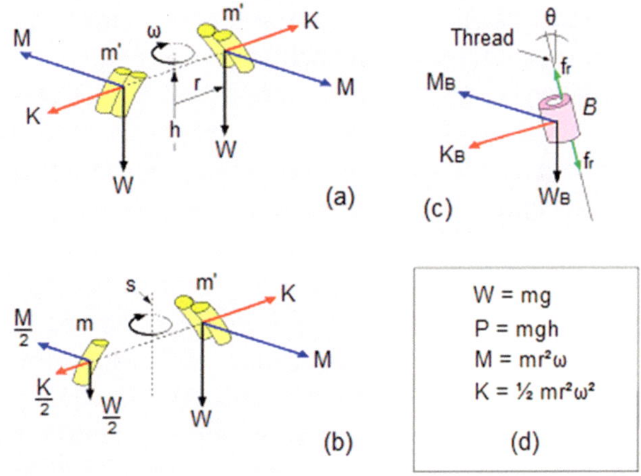

Figure D4. Actions on: the Split key ring and the bead

During the spinning there are four types of energies acting and governing the behavior of the split key ring and the bead. They are the rotational kinetic energy (centrifugal force), represented in Figure D4 as *K*, the angular momentum (*M*), the potential energy or weight of the ring, (*P or W*), and the friction (*fr*). Except for the friction, the other three energies depend on the mass of the ring, and except for the potential energy, the other two depend on the rotational or angular (ω) speed too. See Figure D4d. While the potential energy of the overall ring remains almost constant with time, the other two keep varying and decreasing with time as the spinning movement on the ring is also decreasing.

When the mouth is not included, each sector on the left side of the ring has an identical (mirror) one on the right side at the same distance (*r)* from the rotational axis (*s),* and at any horizontal level (h) from the lowest ring's point, as it is shown in Figure D4a. The rotational kinetic energy (*K*) on each sector are equal but acting in opposite directions, so they neutralize each other. The torque effect of the weight (W) also neutralize each other. On the other hand, the angular momentum is equal of each of these sectors and because they push to keep the ring moving (spinning) on the same direction, what keeps the split key ring on the upright position and spinning is just its angular momentum.

However, when the lighter mouth (*m*) is part of a pair of sectors, it creates at that level an imbalance. The energies (potential, kinetic and angular momentum) on the mirror sector (*m'*) are two times greater than on the mouth (*m*) sector. See Figure D4b. At this instant, both the rotational kinetic energy (*K***)** and the potential energy (*W*) push (down and to the right) trying to make the ring roll to its heavier side. Once the ring turns 180 degrees, the mouth switches position with its mirror sector, switching the rolling effect to the opposite direction too. This switching effect, a very unstable condition, (up and/or down, to the right and/or the left) will continuously happen while the ring is spinning. So, the split key ring mouth should move to a more stable condition, namely to a location where the imbalance is at its minimum.

The friction is on the air contacting the entire ring, and on the contacting spot between the ring and the surface where it rotates. As the time passes by, the friction acts as a catalyst decreasing the rotational speed that promotes all the stages of the split key ring including the switching of position of the mouth and the bead. The rotational speed decreases following a hyperbolic spiral curve (and the angular momentum too). See Section *The hanging spinning wheel* for details.

For the initial high values of spinning speed, the kinetic energy (K) varying with $r^2\omega^2$ (see Figure D4d) prevails over the other two types of energy (*M & W*) and moves the ring mouth close at or to the horizontal ring diameter where the kinetic energy, and the angular momentum are maximum, but where the switching effect is minimum. That is how the first stage (Figure D2) of the spinning ring takes place. The centrifugal force exerted by the kinetic energy, forces and keeps the bead against the ring.

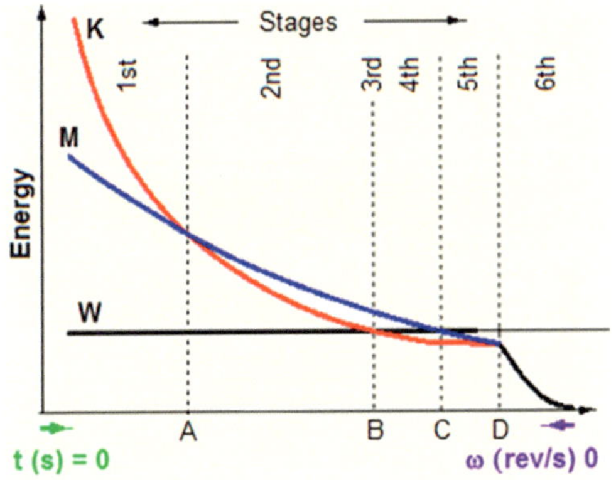

Figure D5. Energies acting on a spinning split key ring

The first stage remains until the spinning decreases to a certain value A. See Figure D5. After instant A, the angular momentum (*M*) prevails over the other two types of energy (*K* & *W*) and the second stage starts. In this stage, the spinning speed is still high enough to generate sufficient kinetic energy that pushes and keeps the bead (*B)* against the thread on an intermediate position or against the ring. Once the spinning decreases to a value B, the bead's potential energy *Wb*, (or its force down component parallel to the thread) prevails over both the bead-thread friction force *fr* and kinetic energy *Kb* (or its force component parallel to the thread), both aiming up. See Figure D4c. The bead falls down to whatever is the bottom (mouth or contra-mouth) section of the ring at that moment. That is how the third stage of the spinning ring takes place, and when the fourth stage begins.

The orbital movement spreads out on the fourth stage and the split key ring keeps revolving on a stable condition until the spinning speed reaches a value, point C, when the potential energy dominates and starts the fifth stage that lasts for a few seconds (few additional complete ring revolutions) when preceded from the previous stages. The sixth and last stage starts when the spinning reaches a value, point D, and the angular momentum cannot keep the split key ring on the upright position anymore. The remaining kinetic energy and angular momentum make the ring wobble on the flat horizontal surface (glass) until its last movement ends.

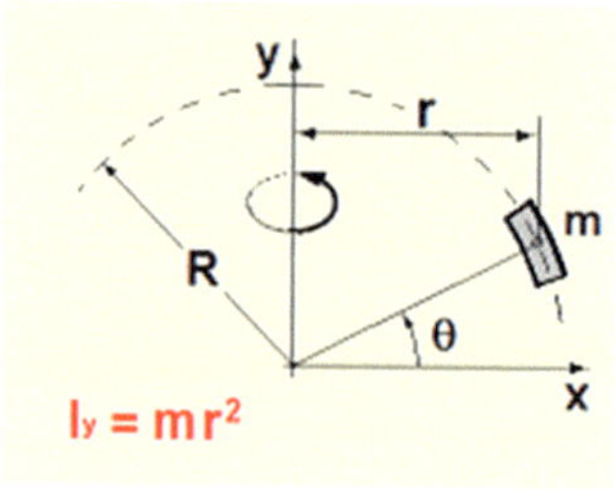

Figure D6. Moment of inertia Vs mouth (spot) location

For a sector of a ring of uniform positive mass *m* spinning at a given rotating speed around *y* axis as it is shown in Figure D6, the term mr^2 of the angular momentum (see Figure D4d) is known as the moment of inertia I_y. That is maximum when *r* is maximum, or when the sector is away from spinning axis Y on the horizontal (left or right) diameter, and is minimum for *r* equal to zero, or when the sector is on the vertical (top or bottom) position. However, if the mass of the sector is "negative" as the split key ring mouth that was made cutting out a small section of one of its two rings, the minimum and maximum values of the moment of inertia switch their positions. For details of this switching see *Moment of inertia of some round bodies*.

The moment of inertia of the split key ring is maximum when the negative or subtracting effect of its mouth is minimum, and vice versa. For having less mass than any other ring pair, the mouth contribution on the energies should be as minimum as possible to allow other pairs to contribute to their maximum. Furthermore, the mouth energy is much less when it is at the ring's bottom where its potential energy is minimal. When the angular momentum prevails, even though the contra-mouth is heavier than the mouth, it goes up as it was observed on the second stage of the key ring spinning. If the ring did not have a mouth, its moment of inertia would be unique or constant and independent of angle θ, and the stages probably would not appear. In other words, this mass imbalance and the spinning generate the other split key ring movements.

For these changing of energies associated to the mouth mass position, we infer that the gravity diameter (or the eccentricity) is a relative measurement of the *mass imbalances* or a *revolving mass imbalance* created by the object revolving about axes passing through its body but not through its center of gravity.

The spinning toy gyroscope

An experiment like the one done with the hanging rotating ring (Part B) was also carried out with a small toy gyroscope centered hanging from a fishing line. In this experiment, it is possible to observe that a spinning (disk) body will generate internal forces when an external action disturbs its spinning plane. When the disturbance is a torque, the body produces two lateral and perpendicular forces. These forces induce an orbital movement and a lifting action or another torque. The lateral movement confirms that there is an excess-reacting energy couple on the horizontal diameter of the toy's wheel. The knowledge gained in the experiments, will allow us to predict the direction of these movements before they begin. It was discovered and validated that a spinning object can simultaneously have two instantaneous centers of gravity.

"Centered" hanged means that when the toy gyroscope gets balanced, its static wheel is on an upright position. Its center of mass is in the midpoint of the length of its horizontal axle when its axle ends are free of any load. When a (metallic washers) load is applied to one end of the axle of the gyroscope, it proportionally inclines its axle and moves its center of mass over its axle to a point between its wheel and the load. See Figure D7a

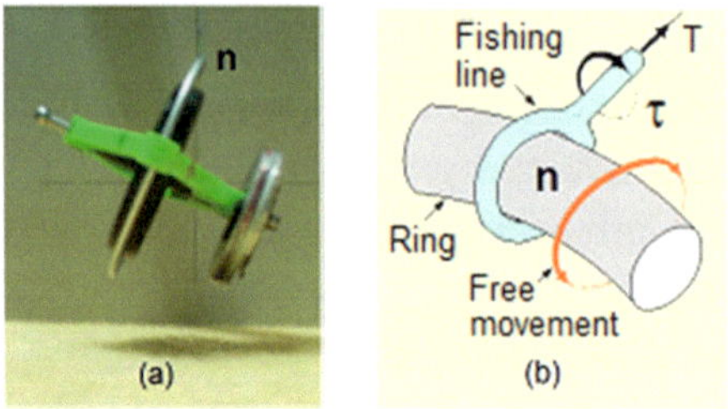

Figure D7. Gyroscope (a) static (b) hanging point

For observation purposes, an axle end and a side of the gyroscope were marked with red and the other side and end, with black (red and green colors respectively in further sketches). The spin of the wheel was done always on the same direction. So, when we are seeing the red axle of the wheel its spinning appears to be on CW[28] direction, and CCW if the end is the black one. The hanging and sliding joint (spot n) in the gyroscope ring, which is concentric with its wheel, is shown in detail in Figure D7b.

The following four types of external disturbances were applied to the spinning gyroscope during the experiment:

- A torque in the hanging fishing line applied on either the CW or CCW direction. The initial torque's magnitude was known in some cases.
- A load was added to one of the two ends of the wheel axle. Two different weights were used.
- A forced vertical (lifting or depress) movement on one end of the axle.
- The torque was combined with one or two of the other disturbances.

---

[28] On a top or front view, CW stands for clockwise, and CCW stands for counterclockwise.

It is worthy to clarify that the torque in the fishing line used on the experiment was variable, alternating, and cyclical. An accumulated torque as the one applied initially, decreases from a maximum value to zero when it is transferred to the wheel, and it is succeeded by an opposite one being accumulated in the hanging element from the wheel, starting from zero to a maximum value. This torque is conveyed to and from the gyroscope through its ring connected to the fishing line.

When a (variable) torque is applied to the centered and hanged, not loaded, and not spinning heavy body, like a "heavy" toy gyroscope, it starts rotating (turning) about an axis parallel to or coinciding with the vertical fishing line and starts to tilt its axle. Contrary to the hanging rotating ring (detailed on section B), its wheel does not reach the horizontal position because the torque is too weak to lift it. On the other hand, when hanging from a fishing line free of accumulated torque, the spinning of the wheel continues as the gyroscope stays in the same position that it gets when it was balanced in the upright or inclined position. However, when one end of its axle is vertically moved (up or down) out of its balanced position, or the fishing line applies a torque to the toy gyroscope, its turning and/or orbiting movements appear. It seems that these movements about a vertical axis passing through the hanging (A) point are latent in the angular momentum of the wheel. In what direction do the movements start?

To answer the above question, let's summarize the experience obtained from the toy gyroscope when it can receive or transfer a variable torque from or to the fishing line.

When it is free of any external torque, the spinning wheel will remain in its static balanced position if the only external force acting on the gyroscope, is the up tension of the fishing line. Its top section, being at a higher altitude than the bottom, has more potential (P) energy than its bottom point. The four end spots that are part of the vertical and horizontal wheel diameters have different levels of potential and kinetic energy. See Figure D8. On this and future sketches the toy gyroscope is represented by only its spinning wheel. In these peripheral spots, the kinetic energy (K) has the same magnitude but different directions, while the magnitude of the always pointing down potential energy (P) is the same in the horizontal diameter and different in the vertical one.

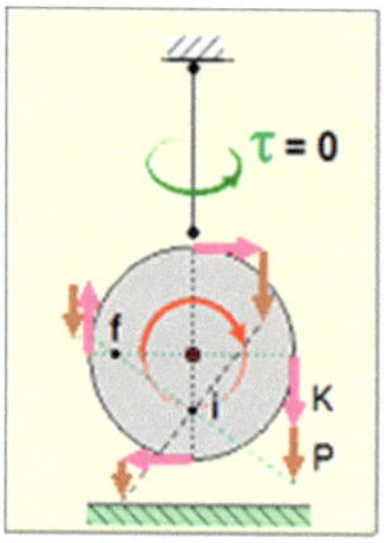

Figure D8. Centered-hanged spinning wheel

Adding these oriented energies[29] ("vectors") in each of such spots, it is obtained the results shown in Figure D8. In the left end of the horizontal diameter, this addition actually is a subtraction of magnitudes because these energies' vectors have opposite (up and down) directions. The direction of the result vectors, not displayed for clarity purpose[30], is practically oriented in the spinning directions, so their net effects promote the spinning. That is the reason why the spinning wheel remains in a fixed position even when there are different levels of energy in these four spots.

The *feeble* or *f point* shown in Figure D8 is an instantaneous center of mass generated by the two (K and P) addition vectors acting in the ends of the horizontal diameter. Similarly, it is possible to find an *impotent* or *i point* in the vertical diameter. Let's define the *feeble and impotent sides* of the spinning wheel as those ones where such points are located. For observation on the experiments it was noticed that the feeble side is on the side where

---

[29] When we add a direction to the scalar energies, they can be considered vectors (magnitude plus direction).

[30] Figure A5a in Section A shows the result vectors for the addition of energies on the two ends of the vertical diameter.

the tangent velocity of the bottom point of the spinning wheel aims at. Notice that the horizontal diameter has the biggest difference of potential and kinetic energies of the four spots in consideration. On its right end, is the strongest concentration of energies; on its left end, the weakest.

As most of the energies in the wheel are concentrated on the strongest spot, this concentration of energy was defined as the *excess of energy* (ΔE) or latent action whose direction aims towards the weakest (f) point. To counteract this excess of energy, an opposite one was defined as the *reacting energy* (ΩE). The energy at the feeble point for being closer to the peripheral point with weakest energy is weaker than the one at impotent point. For that reason, the reacting energy ΩE is located in the weakest or feeble side, and in its opposite side is the excess or strongest energy whose direction has to aim towards the weakest point. The experiment confirmed this assumption.

The wheel feeble and impotent sides where the internal instantaneous (*f* and *i*) centers of gravity are located, are the first ones to sense or feel any external energetic perturbations acting over the gyroscope. This fact was also validated during the experiment. Here is what was observed.

When the toy gyroscope with its wheel spinning is hanged on a fishing line loaded with torsion, as soon as it senses the disturbing (τ) torque, its balance is broken. Some fractions of a second after the toy is released, it simultaneously starts an incipient lateral or orbital movement and lifts a bit the bottom of its wheel, as it is shown on Figure D9a. How these lateral and lifting movements occurred and on what direction? In its incipient releasing, the (τ) torque of the fishing line tries to turn and spin the wheel about the vertical (y) axis but the strong keep *on the same* (actual) *plane effect* (KSPE) of the spinning wheel does not allow this action. However, the fishing line torque keeps pushing, and the toy starts chasing equilibrium. Analogous to a chain that always breaks through its weakest link, the energy is liberated through a spot of lower energy, and/or that yield easily. More than one spot can yield before the chain fails. In the wheel, both its impotent and feeble points meet these conditions. Therefore, the toy moves itself towards the feeble center of gravity spot or in the direction of the excess of energy ΔE, and lifts a bit the bottom side of its wheel. This lifting occurs in a direction that allows the releasing of the fishing line torque.

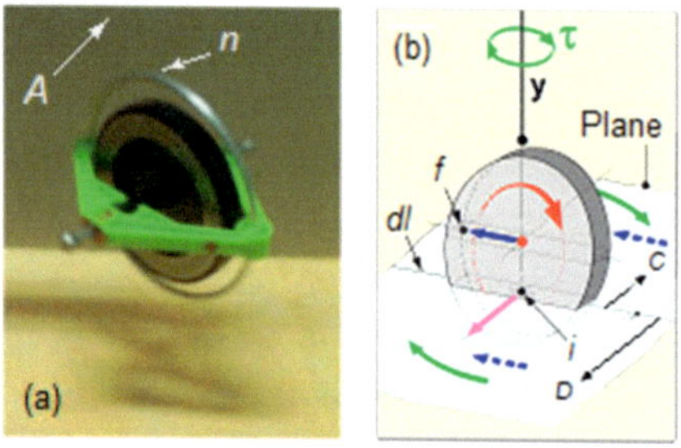

Figure D9. Actions in the spinning toy gyroscope

To determine the direction of this lifting, imagine the toy's wheel crossed by a plane parallel to the lateral force, perpendicular to the wheel's plane, and passing through the impotent spot. See Figure D9b. The plane is divided into two regions (D and C) by the imaginary line (*dl*) that is parallel to the lateral force, and passes through the *i* point. The fishing line torque's effect is perpendicular to the division line when it is crossing or near to this boundary; almost parallel, when away. These "almost parallel" effects are represented in the plane by two curved (green) arrows as it is illustrated in Figure D9b.

Let's suppose that we also make two duplicates of the lateral force (dash arrows in the sketch), and place them on the plane, one on each of its two regions. In only one of these regions, the circular torque effect and the lateral force unite their releasing actions as a couple, and it occurs where one of the force duplicates and the torque "parallel" effect aim in the same direction. So, the red or black axle of the wheel will lift up and move towards this unique side that allows the releasing of the disturbing stimulus of the fishing line. On our example, towards the red end for the spinning and torque direction shown in Figure D9b. The lifting criterion is applicable only in the *switching position* or when the spinning toy is free of any turning movement and the torque in the fishing line is or has reached a maximum value.

The struggle between the turning effect of the torque and the KSPE effect of the wheel remains until the entire torque of the fishing line is completely released. Meanwhile, the wheel starts a lateral movement that should be on a straight line, but because it is tied to the fishing line, the wheel is forced to start orbiting around its hanging (A) point. Notice that when moving in the same turning direction the same axle end is pointing towards the orbiting or turning vertical axis, and the wheel tilts itself towards the horizontal and increases its height. It tells us that the angular momentum of the spinning wheel or its KSPE is a kind of transducer. Which is a device that converts variation of one quantity into those of another (e.g. pressure into voltage). When a disturbing action tries to take the spinning body out of its spinning plane, this effect does something to counteract this action. The angular momentum or the KSPE converts the external torque (τ) into two different actions, namely a lifting force and linear movement of the wheel. Notice that this lifting force (*L*) for acting in impotent point really makes the wheel turn about its x axis; so, it generates the torque (τ_x) about this axis.

During the orbiting of the wheel, its horizontal (x) diameter always remains perpendicular to the vertical *A* axis. Therefore, it appears to be not spinning about the y axis. However, the plane and the axle of the wheel change (turn) position from one place to another, as it is represented by the solid and dashed disks in Figure D10. Thus, on each orbital revolution of the wheel about the *A* axis, the fishing line is twisted one revolution too, and a (τ_A) torsion is also generated about this axis in the hanging and fixed A point. So, τ and τ_A are an action-reaction torque couple acting at the (n and A) ends of the fishing line. The torque of the fishing line is transmitted to the (θ degrees) inclined y axis of the wheel through the n point of the gyroscope's ring. During the orbiting, the junctions at A and n points are acting like a universal mechanical joint[31]. The torques couple seems to acts in the O point located on the *A* axis.

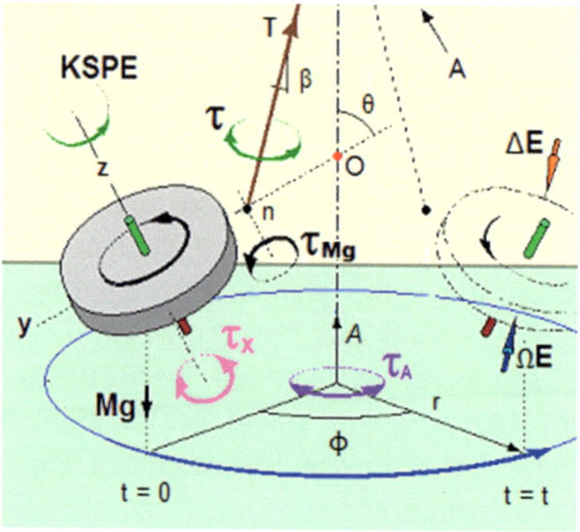

Figure D10. Actions on a centered-hanged spinning wheel

In the experiment, once the movement has started on a particular direction, this direction and the axle end moved up are kept in these conditions until the toy regains the lowest or switching position. To regain this position, the toy gyroscope goes through two stages. On the first or accelerating stage, the fishing line releases its torque from a maximum value to zero. Then, starts the second or decelerating stage, where the fishing line accumulates torque which direction is opposed to its predecessor one. When this new and opposite torque is strong enough, it stops the orbital and/or turning movement. During the releasing stage, the toy gradually absorbs and converts the fishing line torque into

---

[31] A universal joint or universal adapter for socket tools, is a mechanical device to connect two nonaligned spinning shafts. It consists of two U-shape yokes, one on each shaft, connected by a rectangular bar or a cross shaped spider.

a turning momentum that increases its orbital velocity and its height to certain maximum values. These values gradually decrease while the fishing line is accumulating the opposing new torque. Every time the gyroscope reaches the switching position, it shifts both its orbiting speed direction and the axle end to be on the up position. Then it restarts the stage of releasing-accumulating torque in the fishing line. As happened with the hanging split key ring; all these switching (including the accelerating-decelerating process) cease when the angular momenta of the spinning wheel finish and the fishing line is free of accumulated torque. The spinning of the wheel usually finishes first than the releasing of the remaining torque in the fishing line.

The wheel, due to this orbital movement, is moved away from the true vertical (*A* axis) line passing through the hanging point by a centrifugal force generated by its centrifugal kinetic energy. See Figure D10. Neither this force nor the kinetic energy are shown on the sketch. It reaches the farthest or peak distance (r = maximum) from its vertical orbital axis just when the fishing line is free from all accumulated torque ($\tau = 0$). At this moment, the β and θ angles reach their maximum values too, the acceleration process ends, and the decelerating stage starts. The wheel gets the nearest distance (r = 0) when its total angular orbital (τ_A) torque equalizes the maximum torque accumulated, recently acquired or still remaining, in the fishing line during the current set of orbits. In other words when the magnitude of τ equals to the one of τ_A.

During the set of tests detailed on further paragraphs, two external and disturbing effects were simultaneously applied to the spinning wheel. These effects were the torque on the fishing line and a human or "artificial" *elevation* (or depressed) *force* (*E*) on one end of the axle. As the external torque also generates a natural lifting force (*L*) of the axle end, it is possible that such *E* and *L* forces be opposing or promoting each other. They add their effects when acting in the same direction, and subtract when acting on different directions. When the lifting force is opposed and stronger than the depression/elevation's, the spinning wheel almost immediately will switch the up and down positions of the axle ends before starting the turning or orbital movement. This amazing switching of the axle ends will occur whether or not there is a load (see Figure D11b) added to the axle. In those cases, it means that the free end of the axle will also be moved either below or above the level that it got on the (static) balanced position.

The toy gyroscope generates an angular momentum on the direction of its orbital movement that seems to be proportional to the magnitude of the applied disturbance (torque and/or elevation force). In addition to the latent excess of potential energy on its top, the spinning wheel also has a kind of latent momentum associated to its orbital angular momentum that, after a disturbance, is also released into its orbiting movement. This latent momentum also has a big effect on the behavior of the fishing line and the spinning wheel (gyroscope).

On tests with a short and hanging fishing line that has been given 20 twists (turnings), it releases about 18.5 out of these 20 twists on less than one second (0.9s), and the rest (one and a half turns) on about 45 seconds when its bottom end is free, or there is nothing opposing the liberation. As the magnitude of the torsion or torque on the fishing line of a fixed length is proportional to the number of turns or twist given to it, it is safe to say that this twisting number is equivalent to the magnitude of its accumulated torque (τ).

Figure D11a shows the toy gyroscope refrained just after given 20 twists to the fishing line. After that, the toy is ready for either release or, spinning its wheel and release. When the toy gyroscope with no spinning opposes these 20 twists, they make the gyroscope revolve to and fro for an increased total of about 41 revolutions distributed on five sets or groups. That were classified by the revolving direction which is opposed (CW and CCW) in each subsequent set. For one of these tests, from the first to the last set of revolutions, they were, first 30 CW turns, then 5 CCW turns, 3 CW, 2 CCW, and the last 1 CW revolution. In mathematical terms, they can be expressed as just 30, 5, 3, 2, 1 CW-CCW alternate revolutions. Notice the big difference (30 vs 5) between the first and the second set. Each of these groups took 5, 4^+, 5^-, 4, 4^- seconds[32] to complete their revolutions.

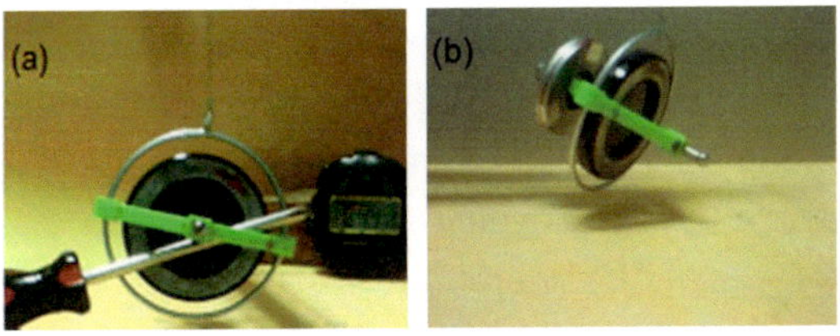

Figure D11. Torque and turning

However, when the wheel is spinning these numbers varied substantially. In this case, they were 29, 16, 11, 7, 5, 3, 2 ... alternate orbital revolutions[33] and their time were 31, 24, 17, 13, 10, 7, 6, 5, 4, 4, 4 ... seconds. Notice now that the difference between the first and second sets of revolutions (29 vs 16) is less than the turning revolutions (30 Vs 5) when the wheel was not spinning. From and including the second quantity, they also change drastically. The decreasing rate between adjacent sets of revolutions and their times is more gradual and smoother, and more sets are needed to reach one or fractions of a turning revolution.

It is worth to notice that when a load (the non-spinning toy) was tied to the end of the fishing line, the first set of (τ_A) orbital turnings, which normally varied from 29 to 31 turns on one

[32] The plus and minus signs mean that the actual time is a bit bigger (+) or lower (-) than the reported one.

[33] At the end, and whether or not the wheel is still spinning; the remaining torque of the fishing line, keeps the wheel swinging back and for on tapering fractions of a revolution for a while. These fractions which oscillation period is almost constant (about 4 seconds each) are not included in the series of data.

direction done by the hanging body, is always greater than the (τ) 20 ones given to the fishing line. It happens because at the moment when the fishing line has just released its 20 turns ($\tau = 0$), the body has also done the same turning, and has acquired an orbital velocity. Due to its inertia (orbital momentum), the body keeps moving, and its turning (orbiting) starts adding an opposing torque to the fishing line that eventually will stop the orbital movement. On its way back, and other subsequent switching of turning directions, the turning quantities in one direction start decreasing because the wheel is losing its spinning due to friction. Most of the friction on the toy gyroscope is generated on its two conical axle ends (no ball bearings) that support the one wheel-axle compact piece to the plastic armature, which also supports the concentric ring.

"The total angular momentum of a system can vary with time only if a net external torque is acting on the system" (Serway and Jewett 317). However, the angular momentum of the wheel is varying with time because its spinning is decreasing with time too.

On another test when a torque on the fishing line and the artificial elevation of one end of the axle are opposite and acting simultaneously, and their difference is small, it is possible to get results like **3**, 34, 20, 13, 8, 6, 3.5, 2, 1.5 ... for the alternate revolutions, and their time to complete each set are **9**, 38, 29, 17, 13, 10, 7^{+}, 7, 6^{+}, 5 ... seconds. These results seem to not agree with the ones shown on previous paragraphs, in which a subsequent value of a series is lower or equal to its precedent. In other words, there was a breaking of the trend of the first two sets of data. However, the difference between the first and the second values of these series of data tell us that the equivalent "elevation" torque was about three or four turnings stronger. It was also possible to infer that initially the disturbing (τ and the elevation) torques were opposed for few (9) seconds. It also means that this torque difference was transferred or added by the wheel to the fishing line in just nine seconds. This late action is reflected on the increment of orbital revolutions of the second group (34) that is above the values (29, 30 or 31) expected for and obtained on the first group of tests. It also confirms the existence of the torque (τ_A) generated by the orbital momentum of the wheel.

The torque (τ_{Mg}) of the wheel weight that always pulls down the partially slant wheel is opposed by a radial torque (τ_X) also generated by the angular momentum of the wheel during the acceleration stage. But during the decelerating stage when the fishing line torque (τ) has changed direction, the radial torque (τ_X) and the weight torque (τ_{Mg}) add their effects. This addition is what brings down both the wheel to its lowest switching position, and the orbital speed to zero.

According to the right-hand rule for torque orientations, the orbiting movements should always aim towards the same direction. However, that is contrary to the results obtained with the gyroscope experiment and summarized in Figure D10. Based on this observation, it is possible to confirm that right-hand rule in some cases does not makes sense at all to define the torque's direction, while the proposed squeezing effect makes more sense. The torques directions aim towards the resisting points (n and A) or towards the feeble point.

The latent momentum also acts like a catalyst that amplifies the peak radius of the orbits. For instance, when the 20-turn torque is applied to the non-spinning wheel, the angle (β) reaches about 2.5 degrees and the angle (θ) about 5 degrees, but when the wheel is spinning or its orbital angular momentum is added, the angle (β) arrives to about 21 degrees and the angle (θ) to about 79 degrees. See Figure D10.

The fishing line could be leaned at an angle (β) from the imaginary and vertical axis (*A*) of the orbital movement, and the plane of the wheel could be inclined (θ) from a vertical plane, that after some time could be displaced at an angle (ϕ) regarding an initial vertical plane. As all these angles, torsions, forces, velocities, and momenta change with time, the mathematics and physics to describe the behavior of the spinning wheel are so complicated, and beyond the knowledge of the writer. However, nature or the energy itself must not only know these variables perfectly, but also what law(s) govern and interconnect all of them. It knows what, went, how, and, even why to modify, adjust, or vary any or all of them to keep the wheel spinning on, or toward a steady condition of minimum energy consumption. That is also true for a bit complicated case, specifically the offset hanged spinning wheel considered in the next section.

In the toy gyroscope experiment, we observed that when a disturbing action tries to take the body out of its spinning plane, the KSPE does something to counteract this action and regain a dynamic equilibrium. It converts the external torque (τ) into two different actions or lateral forces acting on the wheel: one generating its lateral lifting, and the other that promotes its linear and lateral movement. That induces an angular or orbital momentum which maximum value is proportional to the magnitude of the applied disturbance.

The lateral force producing the orbital velocity, is induced as a reaction to an external disturbance, and always aims the same way as the excess energy in the wheel, or from its axis (axle) towards the feeble spot. Direction that is equivalent to the tangent one of the bottom point of the spinning wheel. The lateral lifting force induced at the impotent point and perpendicular to the wheel's plane generating the lifting of one axle end, aims towards the wheel's side where the disturbing torque effect is parallel to and aiming in the same direction as the lateral orbital force. The natural lifting of an axle end at the switching position is associated to both the wheel's spinning and the fishing line torsion directions. When one of them changes direction, the direction of the lateral movement and axle's end going up change accordingly. The existence of an excess-reacting energies couple, and the two instantaneous centers of gravity, allow us predict the direction of such movements.

The offset hanging spinning wheel

On this experiment carried out with a wheel of a bicycle, it is possible to observe that (like the toy gyroscope) it also starts a translation movement when an external disturbing action tries to turn its spinning plane. It was also determined that the force and moment creating the translation are proportional to the magnitude of the disturbance. It was corroborated that the wheel's angular momentum and its angular speed decay following a hyperbolic spiral curve. In this experiment, the disturbing torque (action) has two differences regarding the toy gyroscope experiment; it acts about a horizontal axis instead of vertical, and is almost constant instead of variable, because it is generated by a constant force (wheel weight).

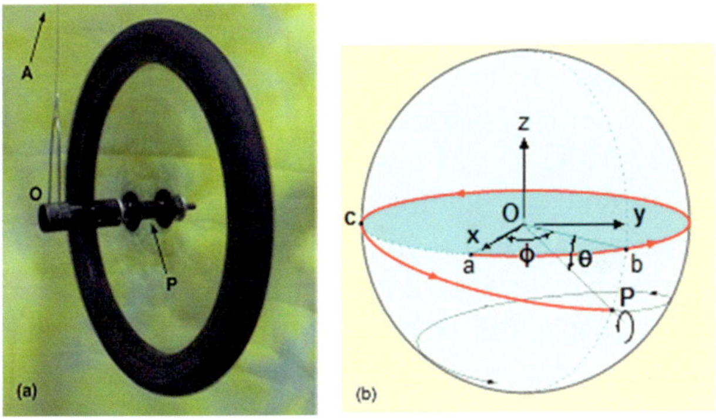

Figure D12: Offset-hanging, spinning-rotating bicycle wheel

Figure D12a shows a bicycle wheel spinning while hanging from a two-leg hook located at one end of its axle that has been extended by a support. The hook acts as a fulcrum which allows the wheel to lean freely from its vertical position when the *hanging point* "O" is offset to its center of gravity (point P). The hook was hanged to a fixed point in three different ways: (1) by a long fishing line as shown in Figure D12a, (2) to a long spring that can withhold tension and torsion as shown in Figure D14 and Figure D17a, and

(3) directly attached to a fixed, strong, and flat metal bar. Of the three hanging arrangements only the spring allows the hanging point to go up and down, the fishing line is the one that offers an insignificant and almost null resistance to a twisting or torsion action, and the hook-metal bar couple has a very strong resistance to torsion and does not allow vertical movement of the hanging point. The offset distance, the type of hanging, and the weight of the wheel were varied to sense how they affect the wheel's performance. On the experiment the bicycle wheel was initially spun to speeds greater than 900 rpm.

For its hanging arrangement type, the spinning wheel is always orbiting about a vertical axis passing through the point A. However in this experiment, when the hanging element does not coincide with such axis, and/or the point O is not in this axis, we call turning the revolving movement instead.

The spinning weight of the wheel includes just the rotating components such as hub, spokes, rim, tire (full of air or full of water), and other miscellaneous components. On the other hand, the overall (static) weight includes also those accessories and components that are neither spinning nor contributing to the angular momentum such as the axle, ball bearings, nuts, hanging hook, support, and devices to measure the revolving speed (rpm) and the tilting angle. The weight of these stationary components is considered to be also acting on the center of gravity or point P of the wheel.

When the spinning wheel, hanging from a long and strong fishing line, is released on the upright orientation, it starts orbiting, describing for a while almost perfect horizontal circles whose center is the hanging point O. See Figure D12b. For spinning speeds over 700 rpm, the descending movement per each orbit is almost nil or hard to observe. Due to the friction of its rolling bearings and air-tire friction, its angular speed and momentum start decreasing. After some while, the wheel starts descending, tilting on each turning until it gets to the position where its axle, the hook, and the hanging line are almost vertically aligned. When the wheel has moved down from the horizontal plane *h,* it has described a small tilting angle θ that is exaggerated in Figure D12b for clarification purposes.

Figure D12b shows the turning-descending trajectory of point *P* of the spinning wheel. The slower the spinning, the faster the descending or tilting rate. At very low spinning speed the hanging point (O) does not remain on one location anymore. It and the wheel, start getting circular turning movements about an imaginary vertical axis passing through the hanging spot but not along the fishing line. Which is not in the vertical position anymore.

When the turning movement is against a very weak resistance, just like air and/or the torsion of a long fishing line, the angular momentum keeps pushing the wheel toward the orbital movement, as if it were enjoying the process of letting go, little by little all the while spinning. The descending of the offset hanged wheel is slow, uninterrupted, and smooth to almost the bottom or horizontal position.

It was observed that the hanging point *O* remained on the same vertical location or at the same height. So, it is deduced that the wheel's overall weight, or a great part of it, is transmitted by its angular momentum (KSPE) action to the hanging point during almost the entire turning descending trajectory. A lateral force (F_T) keeps the wheel orbiting, a vertical up force keeps the wheel in the same plane, and other vertical down force keeps tensed the fishing line. Let's assume that these two vertical forces have the same magnitude (it is validated on the experiment with the wheel hanging from the spring, see details on further paragraphs), and let's name them F_u. They are perpendicular to the (*OP*) line connecting the *O* and *P* points because of its mechanical construction, the wheel is kept perpendicular to the support which is parallel to the OP line.

The spinning plane of the wheel, and the parallel forces (F_u) generated by its angular momentum, have also tilted the same angle θ over the turning vertical *v* plane. To avoid the abruptly falling, the magnitude of these now inclined F_u forces must be greater than the overall weight (F_{wo}) of the wheel. See Figure D13b. Notice that the horizontal components (projections) of these forces cancel each other out because they have opposite directions. Due to the continuous dropping of its angular momentum, the vertical component (projection) of force F_u does not match with the weight of the wheel, that chasing equilibrium descends a bit more to compensate the unbalance. Thus, it is descending softly due to the gradual decreasing of its angular momentum. The orbiting ends when the wheel reaches the horizontal position, or when there is a resisting force (torsion), strong enough to overcome (stop) the turning effect (torque) of its angular momentum.

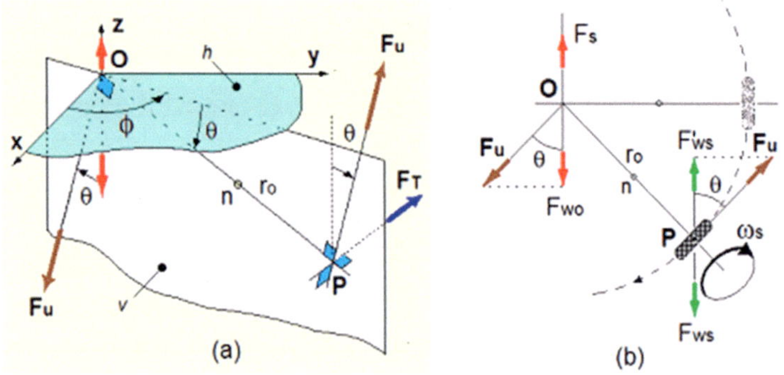

Figure D13. Forces a) upright and turning, b) upright details

In contrast, for a strong resistance like the torsion of a spring, the turning is faster and stops on an intermediate point before reaching close to the bottom. When the turning stops either at the bottom or at an intermediate position, the wheel starts swaying for a while releasing the torsion accumulated on the hanging medium. Once the torsion is liberated, if the still spinning wheel is relocated on the upright position, the turning movement on the horizontal plane is regained on the same initial turning direction.

For the test when the hook is attached to a spring and prior to the spinning up, the wheel was hung on the spring. Once the spring extended its length due to the dead weight of the wheel and got balanced; for reference, the new vertical location or height reached by the hanging point *O,* was marked on a measure tape stuck to a fixed pole parallel to the spring longitudinal axis. Let's call *fix-dot* to this fixed spot on the pole that is at the same height that reached the mobile hanging point *O* when the non-spinning wheel was hanged on the spring. See Figure D14.

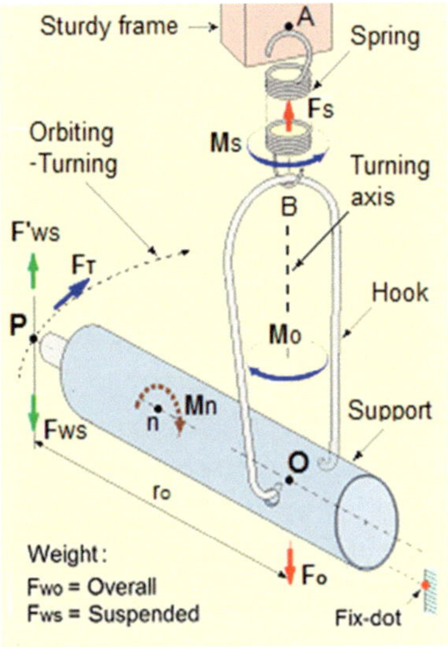

Figure D14. Moments and forces on support device

When the spring extends its length and the hanging point *O* reaches the height of the fix-dot, the magnitude of the pulling up force (F_s) exerted by the spring is equal to the overall weight (F_{WO}) of the wheel. In math terms, $Fs = Fo = Fwo$. Conversely, when the vertical position of the hanging point O is not at the same height as the fix-dot, the down force (Fo) applied on the spot O is not equal to the overall weight of the wheel anymore.

On the experiments hanging the spinning wheel on the spring, the hanging point *O* is free to move up and down. However during most of the turning-descending process, this point remained in a fixed location having the same height as the fix-dot. How is it possible if none of the spring pulling up force reaches the point P due to the hinged (hook) connection, and the wheel spinning was decreasing and the tilting angle of its axle from the horizontal was increasing? Let's first analyze this condition when the wheel axle is still turning over a horizontal plane as shown in Figure D14.

When the hanging point O reaches and remains at the same height as the fix-dot during the orbiting, the down force on the hanging point has to be equal to the overall weight F_{wo} of the wheel. A different force will move this point up or down. Similarly, when the wheel remains in the same horizontal plane while turning (orbiting), its suspended weight[34] (Fws) concentrated on point *P* must be suspended by a vertical and invisible up force (F'_{ws}) for the hanging spot to be a hinged fulcrum. Consequently, the angular momentum of the wheel has simultaneously created two vertical and parallel forces of opposite direction, one *(*F_{wo}*)* on point *O* to counteract the pulling up of the spring and another *(*F'_{ws}*)* on point *P* to suspend the net orbital weight of the wheel. What creates these two forces? It must be a moment, known as *moment of a couple*[35]. This moment, assuming the support is weightless, is equivalent to applying a twist or torque M_n on intermediate (*n*) spot of the imaginary line connecting point *O* and point *P* or about the midpoint of the offset distance r_o.

To visualize the effect of this moment of a couple, let's suppose that we have the arrangement shown in Figure D14, but

---

[34] In this position, the suspended weight of the wheel does not include the weight of the hook.

[35] Two non-collinear and parallel forces having the same magnitude, and opposite direction form a torque known as a moment of a couple. They tend to make the body rotate.

the wheel is not spinning nor orbiting, meaning it does not have angular momentum and the moment M_n does not exist. If we apply enough torque (similar to the one applied to a nut of a car wheel) with an x-shaped lug nut wrench on this intermediate *n* point, we should be able to extend the spring to the same location obtained during the test and also keep the wheel suspended with the support on the horizontal position. That is, applying a torque (twist) on an intermediate point of the support we would generate the two forces (F'_{ws} and F_{wo}) described on the previous paragraph.

Because the spring has one end fixed against a sturdy frame, it is able and can withstand torsion, whose magnitude is proportional to the turning angle (ϕ) that it has been twisted or turned[36]. This resistant twist (*Ms*) of the spring observed on the test is proof of the existence of both the turning force F_T and its moment M_o around the spring longitudinal and vertical axis passing through point *O*. As the turning increases, the resistance becomes stronger, and if the rpm has dropped to a value at which the moment (M_o) created by the turning force F_T cannot overcome this resisting torsion anymore. After that, the turning stops, the wheel swings down and all the torsion accumulated on the spring is abruptly released through the wheel axle-support. That starts revolving fast and backwards to the turning direction or on the same direction as the spinning.

It is worthy to say that the Figure D14 and Figure D13 apply for the three resisting torsion levels tested during the experiment. The elements and their resistant levels were: fishing line (the lowest), the spring (middle), and hook (the strongest resistance).

In experiments with the hook directly fixed against a sturdy frame, the energy (angular momentum) seems to be alive, because it creates a turning (orbiting) torque that struggles against the big resistance (torsion) generated by the strong hanging element. When it realizes that cannot defeat the resistance, as it does

---

[36] This known fact was validated on twisting tests performed on the hanging spring.

against the spring, it relinquishes the struggle and almost instantly (in less than five seconds) moves the wheel toward the horizontal position, a more stable or less stressing condition, where it can release its spinning energy. For an offset of 65 mm and the bicycle wheel of 3.6 kg, its turning torsion almost destroyed a 3.175 mm (1/8") steel bar hook that was disengaged from the support, releasing the wheel that was still whirling at high speed. It was not a pleasant situation. For an offset of 93 mm, a stronger hook made of steel bar of 4.76 mm (3/16") was also elastically deformed and twisted. Due to the magnitude of this deformation, these tests were not attempted for an offset of 151 mm. Similar results, but not as dramatic as the ones obtained for a bicycle wheel were attained from a toy gyroscope when it was both offset and center hanged from a metal wire of 1.6 mm (1/16") that offers to the toy an almost infinite resistance to the torsion.

In summary, hanging from a spring, the spinning of the hanging wheel creates two action-reaction couples acting simultaneously over its support (see Figure D14 and Figure D13). One is the turning movement which force F_T generates a horizontal moment M_o (action) around the hanging point that is counteracted by the spring torsion (reaction). The other couple actually are two pair of forces generating the vertical *moment of a couple*, M_n, namely one pair acting over point *O* (force F_s and its reaction keeping the spring end on the same vertical position) and the other is the wheel weight (action) and the force that keeps it suspended (reaction). The two moments (M_o and M_n) generated by the angular momentum are perpendicular to the wheel spinning plane and are oriented 90 degrees apart from each other.

From several tests, it was observed that the turning direction is always opposite to the movement direction of the top of the upright wheel. What makes it behave in this way? Three possible effects promoting this behavior were visualized. Figure D15a shows a cross view of the support on the surrounding of the hanging point O. When the turning is on the clockwise direction, the whole weight of the wheel is practically withheld on just the right leg of the hook. Resting on just one dot that is offset from the (O) center of mass of the wheel's support, is a very unstable condition. Trying to regain a more stable condition, the system moves the wheel towards its center of gravity, or to the left, that is the direction taken by the turning movement. The second contributing effect is by the friction on the bearing balls enclosed by the wheel's hub and the bearing cone. Only the top balls of the ball bearings are actively carrying the wheel's weight. See the cross view of Figure D15b. The bottom of these top-rolling balls has most of the friction that opposes the rotating direction. The third effect, the one that really induced the orbital movement, is generated even on a media with no friction at all, and is a kind of "spinning" action, labeled as *excess of energy* that is explained in detail in section *The spinning toy gyroscope*.

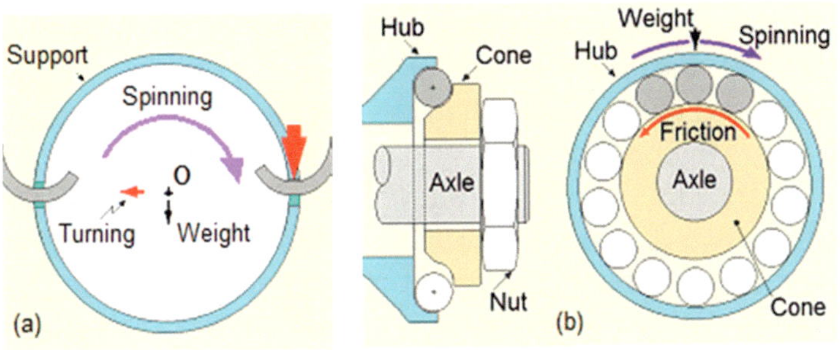

Figure D15. Wheel: a) an unbalanced hook, b) ball bearings

When there is no offset, or when the hanging point is aligned with the center of gravity (point P) of the upright spinning wheel, there is not a steady turning movement in any direction. Instead, the wheel turns a few degrees to one side and then returns, and then keeps going back and forth between these ends, or the turn just reaches and remains in one position until the spinning is very slow or ceases completely. These were the same results obtained on the experiments carried on using the toy gyroscope. Therefore, it is possible to state that the lateral (turning or orbiting) movement along the spinning plane and the moment whose turning effect is parallel to such plane, are generated when other external energy (hanging offset, torque, or force) disturbs or takes out of equilibrium the spinning wheel, toy gyroscope, or in general a spinning object.

According to the results of the experiments done for different angular momenta (wheel of same geometry but different weight) and different torsion resistance (fishing line and spring), the torque (M_o) on the spring is 17 or more times smaller than the torque (M_n) for a hanging offset distance. For three revolutions in the same horizontal plane, the M_o in kilogram-meter (kg.m) was about 0.0069 Kg.m, while M_n for two offset distances varied from about 0.12 to 0.32 Kg.m. However, greater this distance (or disturbing energy) the faster their acceleration (the orbital revolutions occur in less time). For instance, the first revolution, from the short to the long offset (65, 93, and 151 millimeters), took the following average time in seconds (19, 14, and 9). These results show that the torque that promotes the linear or orbital movement is much smaller than the one that forces the wheel to stay in its orbiting plane.

Figure D16 compares the profile of the rotational speed (rpm) for two different net weights (1.668 kg and 3.556 kg) of a bicycle wheel. When the actual revolutions per minute (rpm) measured during the test were compared against mathematical curves, their decay rates closely follow hyperbolic spiral curves (see *The circle and the hyperbolic curve* for details about this curve). As expected, neither the tilting nor the resistance to turning impacts their decreasing rates. For having a constant moment of inertia (same mass, shape, and spinning axis), the angular momentum (defined as the moment of inertia times the rotational speed) of a wheel depends only of its speed. For that reason, its angular momentum should have the same time-decay-profile as the one of the rotational speed, or a hyperbolic-spiral profile.

Figure D16 also confirms what we know by our day-to-day experience. It is much harder to stop a moving truck than a flying mosquito. The heavy wheel will remain moving more time than the lighter, and has a flatter spinning decay rate. The greater the angular momentum, the lower and flatter its decay rate. The Earth is a massive object revolving at very low speed on a medium (outer space "vacuum") of very low resistance to the movement, so its angular momentum is constant in our low timescale.

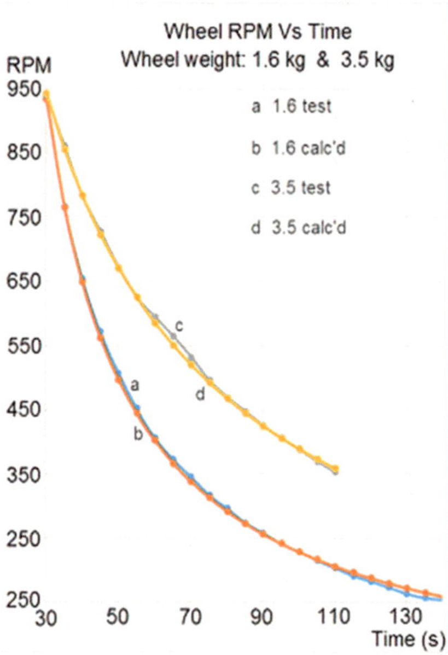

Kg	1.668		3.556	
Time (s)	Case			
	a	b	c	d
	RPM			
30.4	933	933	938	942
35.4	766	764	860	855
40.4	654	647	782	783
45.4	571	561	727	722
50.4	507	496	671	670
55.4	452	444	625	624
60.4	406	401	595	585
65.4	373	367	564	550
70.4	345	337	531	519
75.4	317	312	495	492
80.4	296	291	469	467
85.4	274	272	447	444
90.4	258	256	425	424
95.4	241	241	406	406
100.4	229	228	388	389
105.4	215	216	369	373
110.4	204	206	352	358
115.4	192	196		
120.4	183	188		
125.4	174	180		
130.4	164	172		
135.4	157	166		
140.4	155	159		

Figure D16. RPM comparison for two wheel weights

When the spinning (RPM) speed starts decreasing, the wheel starts tilting from its vertical hanging element. Figure D17 shows the devices utilized to measure the tilting angle of the wheel for the three hanging offsets. For low spinning speeds, the tilting has increased several degrees and the support hanging and moving point O starts to orbit around the fixed hanging point. However, the orbital movement is about a small circle. Figure D17a illustrates this situation for one test. We have named θ this angle but regarding Figure D10 for the toy gyroscope, actually it equals the difference between θ angle and the small β angle. Based on these observations, it is possible to state that the tilting angle is about the true vertical line.

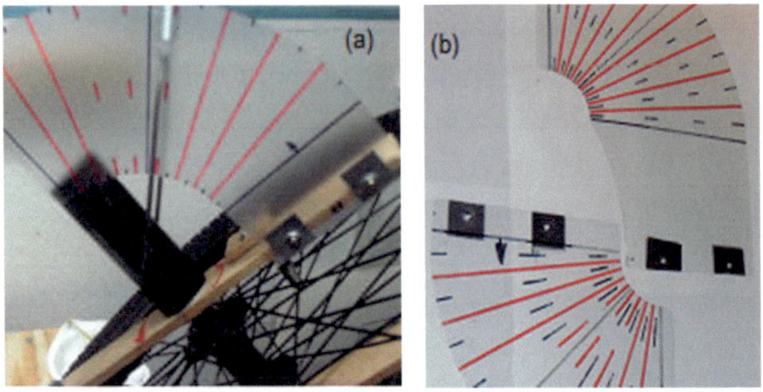

Figure D17. Device to measure tilting angle

Figure D18 summarizes the turning (orbiting) angle (ϕ), tilting angle (θ), and rotational speed (ω) of a spinning bicycle wheel against time for a low and high (fishing line and spring) resisting torsion, and for two hanging offset (65 mm and 151 mm) distances. The ϕ angle is given in revolutions of 360 degrees, while the θ angle is in degrees. It is observed from the graphs that:

- The tilting is proportional to the offset. The higher (lower) the offset, the faster (slower) the tilting rate.
- The tilting is proportional to the resistance against the turning. The higher (lower) the resistance to turning, the faster (slower) the tilting rate.
- The turning is proportional to the offset. The higher (lower) the offset, the faster (slower) the turning rate.
- For a given offset, the turning rate of a wheel just depends on its angular momentum (rpm for a fixed mass) and is independent of the resistance against the turning movement.

The effect listed on the second bullet point above is magnified on experiments with the hook directly fixed against a sturdy frame.

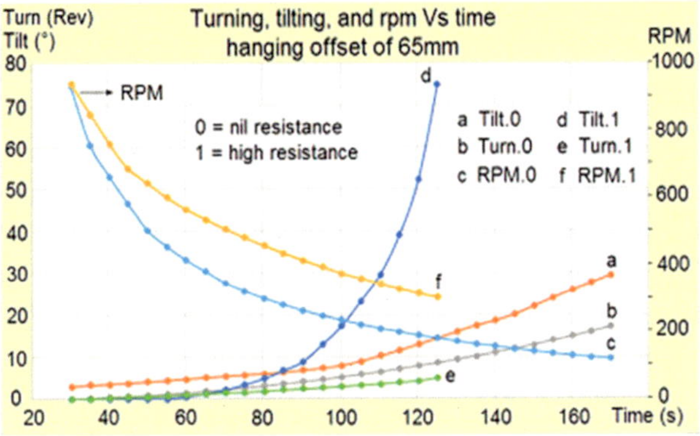

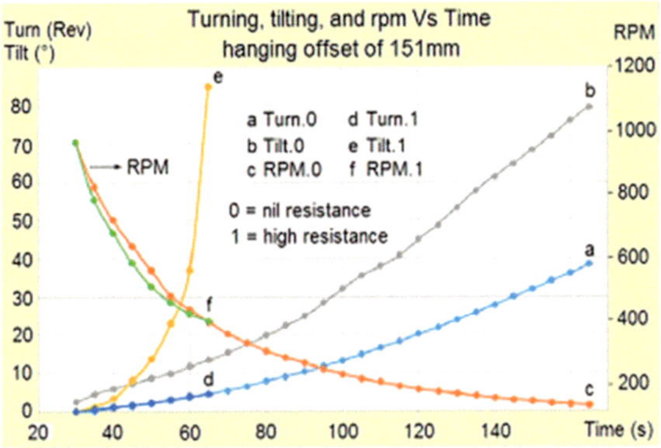

Figure D18. Turning, tilting, and rpm of a spinning wheel

We have seen that the two moments (M_o and M_n) generated by the wheel's angular momentum, or its KSPE, can happen simultaneously as a pair, or one of them can exist without the other for certain conditions. From Figure D18, it is conveyed that if the tilting effect, gravity is nil, and the resistance to turning is so small, the turning will almost always happen over just one plane (no tilting) and at a very slow or constant (flat) rate. Can the physical medium, the support, interconnecting the hanging point and the spinning object be necessary if the tilting effect (gravity) disappears, or is very small? Can the resistance to turning disappear completely? It seems that our wheel must overcome the air friction created by its traveling, even at a very slow velocity.

It is possible and feasible that there is something, or a place where there is no resistance at all? Is there no resistance at all in outer space? Is a rocket propulsion just an interchange (conservation of energy) of only momentum? Do the exhausted gases not have an actual contact (reaction) against the rocket wall? Is there no kicking of the exhausted gases against the outer space, which is "full" of vacuum? We are not able to see the vacuum but it has recently detected as a spatial energy that does not predominate in any direction.

Traits of air-suspended spinning balls

The next experiments were inspired by well-known ones done with pin-pong balls, (Spangler 11) and are intended to demonstrate that a flux of energy (up-stream of air) and friction can induce rotation on a spherical object like a ball or the Earth.

Due to the fabrication process, in most cases a ball is not homogeneous or uniform. So, a real ball would most likely have an internal spot that could be heavier or lighter than the rest of the mass of the ball. The unique ball's diameter that goes through such point and the center of gravity was defined as the *gravity diameter*. Whose peripheral ends were painted in two different color spots, or marked with a circle and a cross to stand them out from the balls' color. On further paragraphs, we make reference to these peripheral spots as the *heavy* and the *light spots* even when they might have the same weight (mass) as any other external ball spot.

Two procedures were used to find such diameter, let's first explain the one that works for light and/or hollow with uniform thickness and smooth surface balls. For spherical balls resting on a smooth and leveled flat surface (glass), the gravity diameter is aligned vertically by gravity, and the heavy spot will be close to the diameter bottom end, and the light to its top. In this case, a ball is tossed several times on a flat glass, and for each toss the most upright point of it is identified and marked. Three scenarios are possible for the aim-up points: (a) they are localized on a narrow spot, it means the ball eccentricity is big, (b) they reach random locations (low eccentricity), always located on one of the two semi spherical sides of the ball. In these two cases, the upright dot was selected close and among those with more aim-up, and (c) they are randomly distributed over the entire surface of the ball. In this unlikely case the gravity and geometry centers should either coincide, or be very near. The second procedure consists of placing on resting water, heavy (big) balls that float on it, and mark the spot that always aim up. This method is very accurate, but it does not work on light weight balls because the "sticky" surface tension of the water is greater than the tiny turning moment created by any weight irregularity on their thickness. They just sink a bit and their heavy spot does not always reach the bottom.

Once the gravity diameter was identified on the small balls, the heavy and the upright spots were painted in two different color dots that stand out with the balls color. See the two small balls in Figure D20. On the "big" ones these spots were marked with a circle and a cross. The spinning and rotation of the balls described in further paragraphs are related to the gravity diameter.

Friction, resistance a body encounters in moving over another, has been disparaged and disguised through the years because it is not easy, especially in outer space motion, to see, feel, or calculate it. It is also present inside matter, and it is known by different names, like viscosity (internal friction in fluids), surface tension (intermolecular cohesive forces), drag (frictional force between a solid and a fluid), but it is much more that. It is energy that also promotes and catalyzes the energetic equilibrium. According to scientists, the Earth's rotational speed is slowing slightly with time. The length of a day was about 21 hours at 600 million years ago to the current 24 hours value (Wikipedia, Earth's Rotation - Rotational Velocity, 2nd paragraph). If that data were reliable, what is stopping the Earth? Is it friction?

To study the importance of the friction on the equilibrium of energy, homemade experiments were carried out using a group of balls of different diameters, weight, and surface finish. To create a vertical-up stream of air to lift and keep "floating" the balls, a common two-speed hair drier, and a leaf blower were used. Long pipes were attached to the hair drier and the blower to get one uniform and homogeneous flow of air. Figure D19 summarizes some features of these components. The weight reported is the heavier plate that the accelerated flow of air can tilt to escape when the plate is located on the top edge of the vertical pipe. The weights of these plates are much heavier than the weights of the balls that the air stream can lift and keeps floating. Compare tables of Figure D19 and Figure D20.

Feature	Hair drier speed		Leaf Blower
	Low	High	
Weight[(1)] (gr)	60	140	2600
Power (watt)	218	380	1496
Height[(2)] (mm)	780	1140	>>1900
Pipe ID[(3)] (mm)	42		77

(1) Lifting a flat object; (2) reached by a fabric thread lifted by the air stream (3) internal diameter

Figure D19. The hair drier and the leaf blower

Ball / Features	Earth globe	Ping pong	Kid tennis	Practice Golf
Weight (gr)	24	2	45	2.8
Diam. (mm)	74	39.8	64	40
Surface	smooth		rough	
Body	solid	hollow		solid

Figure D20. Characteristics of four balls

When a ball is placed on the air stream three scenarios are possible: (1) it does not lift either because it is so heavy, and/or its volume (or projected area) is so small; (2) it is blasted off, out and away of the air stream because it is too light, and/or the energy of the air stream is so strong, and (3) it remains "floating" or suspended on the air. The experiments explained on further paragraphs, are those dealing with balls that remain "floating" on the air stream. See Figure D20 for data of four different balls used on such experiment.

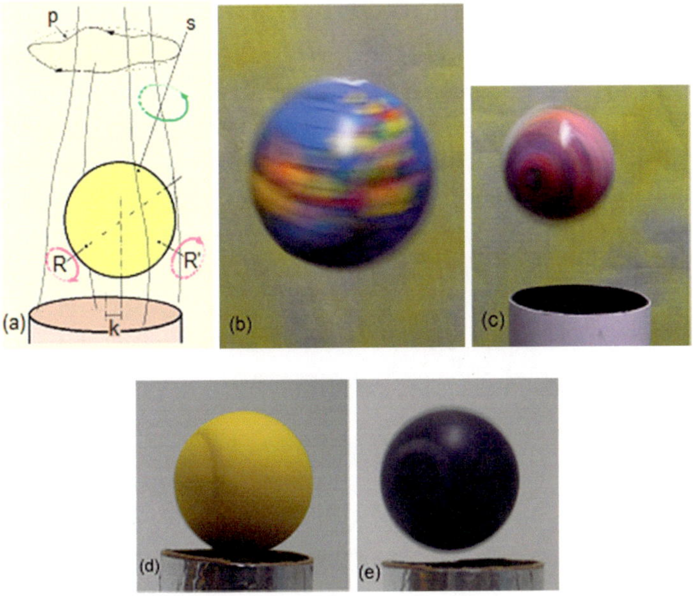

Figure D21. Balls floating and spinning

Figure D21 shows pictures of balls spinning and "floating" in the air. The two at the top were obtained using the leaf blower, and the others using the hair drier at its low speed. Most of the big and heavy balls floated with none or few spinning about an almost vertical *s* axis following a waving and wobbling[37] orbital *P* path as the one shown in Figure D21a. Others, like the one in Figure D21b, followed a *P* path with nutation[38] movements. Some, like the one in Figure D21c, do not start spinning by themselves. When they are forced to start spinning about a vertical axis, they keep this orientation for a while, but as soon as they reach a high rotating speed, they switch their rotational axis to a horizontal position.

On the contrary, the small ones like a furred ball, (see Figure D21d), and a smooth ping pong ball (see Figure D21e), when placed on the air stream, initially float with none or few spinning. Then due to the air-ball skin friction, they start getting rotational movement and accelerate to a peak spinning on a horizontal plane perpendicular to the main air stream flow.

In general, the behavior of the small balls is very similar, except for two important differences. The one with a rough surface keeps the rotational axis on an almost static location on the horizontal plane (it means no rotation about *s* axis). The other with a smooth surface keeps the rotational axis on a position *R* for a while, and then slowly moves it to the other location *R′* on the same horizontal plane. Therefore, two spins occur simultaneously, one at high speed about a horizontal axis, and another at low speed about a vertical axis.

The predominant spinning of most of the smooth balls is not quite uniform. It fluctuates and reaches a steadier condition at high speed. On the other hand, the spinning of most of the rough ones is almost uniform and constant, reaching high values that are sustained for a long time. Although both oscillate in the vertical direction, the oscillation of the rough (golf-practice) balls is almost nil for both air drier speeds, while for the smooth (Ping-Pong) ones this movement is more evident, especially at the drier high speed when the oscillation varies from zero (ball touching the cylinder and then jumping) to about six times its diameter. During a high

---

[37] Changing position on a random way.

[38] Wavy and oscillation of a spinning object.

range of vertical oscillation, their rotation is almost nil, but once it starts, the oscillation range is low and the balls and their spinning axes keep a position for a while.

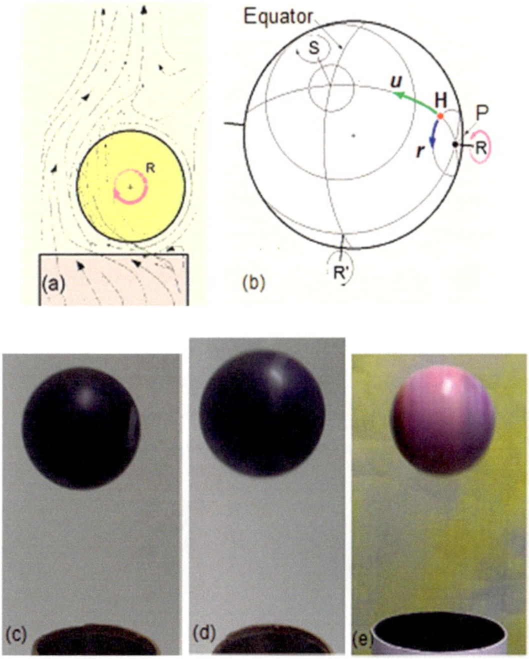

Figure D22. Balls floating and spinning about a horizontal axis.

Balls having a diameter a bit smaller than the internal diameter of the pipe spin about a horizontal axis and lift up approximately six times its diameter above the pipe outlet edge. Conversely, balls having a diameter much bigger than the pipe's, are lifted close to the pipe outlet. As they do not have enough room to be horizontally moved toward one side, they start moving back and forth from one side to another of the pipe rim. They eventually get out of the air stream and fall or start orbiting and wobbling following a *P* path as shown in Figure D21a.

Similar to the contra-mouth in the split key ring spinning on one of its vertical diameters, the heavy spot (*H*) of the gravity diameter of a ball keeps chasing for equilibrium and/or stable condition. See Figure D22b. This *H* spot spinning about a horizontal axis keeps moving on the *u* direction, towards the *p* pole or towards the equator depending on the spinning value. Once on the other side, it restarts its movement on the opposite direction. Figure D22c and Figure D22d are snapshots of this process. Is this switching process perennial? No, because the interaction between the air stream and a ball is not perfect, and the spinning varies. When it reaches very high values, the spot velocity on the *r* direction overcomes the one on the *u* direction, stopping the switching process. The tennis ball (Figure D22e) reaches this condition and keeps it for a while due to its high rotational momentum. The switching process also stops when the air vertical pushing is momentarily lost (the air stream is neither uniform nor stable) and the ball falls down. For some range of spinning speed, the *H* point keeps rotating most of the time between a pole and the equator without reaching none of them.

Features		Ball	
		Red	"Earth"
Diameter at seam (mm)		103-104	72 - 73
Diameter at "poles" (mm)		102	74
Weight (gr)		100	24
Eccentricity[1] (mm)		2[2]	3.1[3]
Heavy dot at top	Spin (RPM)	35	35
	Tilting[4] (°)	8	48
Heavy dot at bottom	Spin (RPM)	52	>200[5]
	Tilting[4] (°)	6	54.5

Notes:
(1) Geometric center to Gravity center distance
(2) Gravity axis perpendicular to ball's seam plane
(3) Gravity axis about 8.5 mm from ball's seam
(4) Angle between vertical axis and gravity axis
(5) At quasi-stable condition: 425 to 550 RPM

(a)

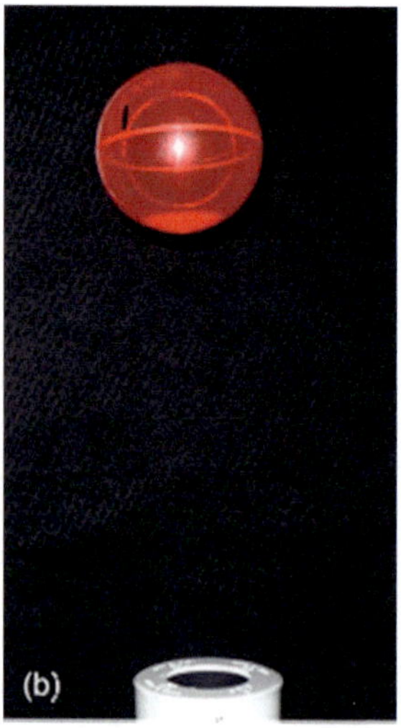

Figure D23. Traits of balls spinning about a "vertical" axis

Figure D23 shows the traits of two balls and some results obtained when they were floating in the air stream and spinning around an axis a bit tilt from the vertical. On the one labeled "earth ball" (see Figure D20) and shown spinning in Figure D21b, the light spot is on the coast of "Antarctica" facing the "Indian Ocean" and about 8 mm away from its seam which goes through its actual South Pole. The heavy spot of this ball is located on the north side of the Canada sketch. The second, a *red*, hollow, and bigger ball which light and heavy spots resemble the poles of our planet because they are part of the hemisphere sides joined by a seam also resembling the Earth's equator. The heavy spot of it, or one of its "poles," is clearly identified for an accumulation of material forming a small and solid spherical cap. See the ball's bottom in Figure D23b.

When located on the air stream coming from the 77 mm (3") diameter pipe, the red ball was barely lifted and started jumping to and fro to different points of the pipe edge until eventually it was out the air stream and fell without spinning at all. To make it lift and spin, the pipe exhaust end was covered with caps drilled with holes of diameter varying from 25.4 mm to 50.8 mm (1 in to 2 in).

For this range of air stream diameter[39], the spinning values were close to those shown on table of Figure D23, and the lifting of the ball was about 20 mm (0.78 in) for a cap hole of 25.4 mm (1 in) and higher heights for bigger diameters. The mean ratio of air jet diameter to ball diameter was 1.04 (77/74) and unique for the earth ball while for the hollow red one it varied from 0.24 (25.4/103) to 0.49 (50.8/103).

When big balls were placed on the air vertical stream, some were so "heavy" that they did not even lift from the pipe edge, and others only oscillated from one spot to other on the edge of the pipe with no rotation at all. To get them spinning, "artificial" protuberances were attached to their surface near or around the

[39] Taken the hole diameter instead of the stream contracted's

gravity axis interconnecting the heavy spot of the ball and its diametrical opposite ("lighter") side. Among the protuberances used were just a scotch tape or flexible fins as it is shown in Figure D24. The fins in some cases either was just one located near to the lighter spot, or when two, the additional one on the equator, or both located symmetrical to the lighter spot among other arrangements.

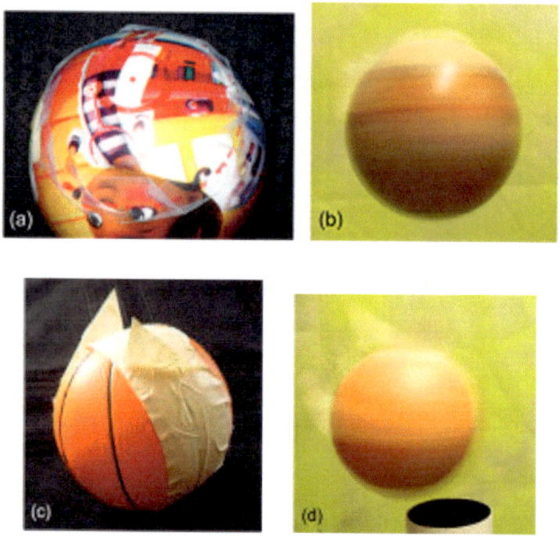

Figure D24. Balls with artificial protuberances

After the modification, these balls spun at very high and sustained velocity, about an axis a bit tilted from the vertical, position and inclination that they kept for a very long time. A "comic" ball that weighs 54 gr and has a diameter of about 120 mm (4.72 in) is shown spinning in Figure D24b, and a "solid" "basket" ball that weighs 90 gr and has a diameter of about 118 mm (4.68 in), is shown spinning in Figure D24d. On these balls, their protuberances were always in the top half as it is suggested by the phantom images of the spinning balls of Figure D24. They did not shift as the balls in Figure D23 did. Regarding stabilization, these balls are much more stable than the ones without protuberance when they are spinning on their more stable arrangement or having their heavier half on the top.

The balls on the energized streams (air and water) stand out a well-known but ignored principle or fact, namely that the interchange of flux (quantity by unit of area) of energy shall occur gradually and at levels of the same order of magnitude. When this difference is big, the most high level will entirely absorb or minimize the one with the low level or will blast it out. For instance, when the air stream (kinetic) energy is much greater than the (potential) energy of the ball, as soon it is placed on the air stream, it skyrockets away from the stream. However, when the energetic levels are close enough, the ball will float or be suspended on the air stream. We are not blasted away (incinerated) by the sun because we only get from it enough energy to keep both the Earth spinning and live in it.

A solid sphere can be visualized as a huge group of spherical shells having a common geometric center, like a group of ping pong balls of different sizes assembled concentrically as a whole. Assume that the center of gravity of the sphere (all the shells together) does not match with the geometric center (that is the most likely situation), and assume that each of these spherical shells has uniform thickness and uniform density except for one and only one that has one spot lighter or heavier than any other point of the sphere. This spot and the center of gravity of the sphere are located on this particular shell. Let's call it, *the gravity shell*. On the gravity shell there shall be only one axis, specifically gravity axis, touching the center of gravity and the geometric center of the sphere (the gravity diameter is co-linear with this axis, see *Balancing and the Add-cut theorem* for details).

If a slice is obtained by cutting the gravity sphere by two parallel planes that are also parallel to the gravity axis, and each plane is at the same distance of this axis, a ring is generated. Let's name it *the gravity ring*. For having uniform thickness, and uniform density except for one spot, the gravity ring resembles the split key ring. Due to this exceptional spot, it is expected that the spinning behavior of a sphere shall be very similar to its gravity-ring behavior, also very similar to the split key ring's. For analogy with the split key ring, the gravity diameter (or the eccentricity) of air-suspended balls is a relative measurement of the *revolving mass imbalance* created by the ball revolving about axes passing through its body but not through its center of gravity. The experiments with the balls and the split key ring collaborate this assumption.

Creating vacuum

On this section the experiments are designed to create vacuum without using mechanical devices (vacuum pump, vacuum cleaner, etc.). We will make water "float" in the air, and make an inflated balloon submerged in water keep its shape by an external vacuum action when the filling air escapes. Using these tests we corroborate that pressure, vacuum and adhesive forces can coexist in a medium.

Among several old and well-known homemade experiments for kids dealing with air, the procedure of one of them, with some variations, is detailed in these paragraphs. To carry it out, use a rigid and transparent receptacle having a fixed internal volume like a drinking glass, or a vase with a flat edge, and fill it with water without reaching its brim. Next, wet the entire brim and place and cover it with a piece of flexible and plastic film (or a piece of cardboard). Press it slightly with the palm of one of your hands. Then, keep your hand over the film and carefully and fast turn the assembly upside down over a basin. After that, take the hand away from the film. If everything goes well, you are good to go. See Figure D25a.

The vase retains water in its interior even when it is upside-down! It happens even for a column of water that only reaches a few millimeters (fraction of an inch), or if the vase is full of water, provided that on the whole brim of the vase there is a layer of water. This layer creates a seal between the plastic film and the vase making its interior a closed volume which can vary by the deformation of the film. See Figure D25b.

If during the turning over process no water escapes and no air gets in, the only thing that changes is the vessel orientation.

Thus, on the upside-down position, the water weight bulges down the plastic film creating an additional internal space for the same volume (mass) of water and the air. In this condition, they are stretched out, or drawn out into extension or size, or under vacuum[40]. That is what we want to do when we suck a liquid through a straw. We really want to stretch out the liquid.

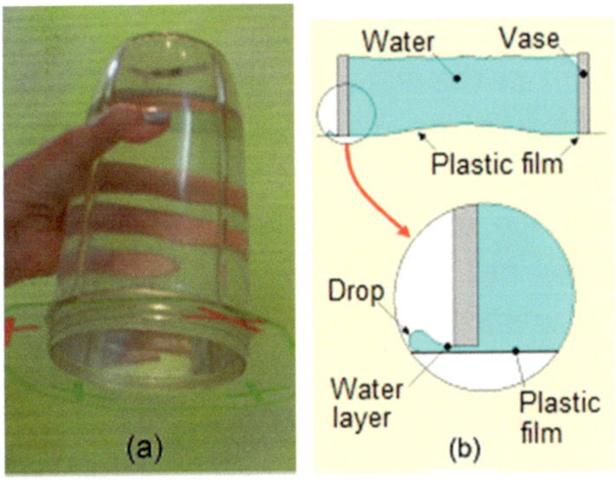

Figure D25. Vessel under internal vacuum

When some water gets (as may occur) out the vase, part or drops of this water may be sucked back into the vase again. This sucking back effect is an evidence of vacuum, the same created by a vacuum cleaner. In this case, we have the same closed volume, but less mass of water which sucks and bulges up the film to compensate the lost mass (volume) of water.

Another way to check out the created vacuum, instead of using a "thick" plastic film (0.008" = 0.203 mm thick) is to use a piece of plastic bag that is very thin (0.0005" = 0.0127 mm). To ease the turning, take care of placing a flat and rigid piece over this film to avoid it getting in or out the level of the brim of the vase. Do not allow any water between the film and the flat piece

---

[40] Two definitions of vacuum: space or vessel from which air has been partly or completely removed; decrease of pressure below normal atmospheric value. (Sykes and J.B.)"

because by the surface tension of this sandwiched water, when you take away your hand and/or the flat piece, it could also drag away the bag film. On this version of the experiment much water usually gets out of the vessel and the curvature formed by the water-film pair got a very visible spherical cap shape. That could be greater than the one sketched in Figure D25b. The size of the visible curvature of the film (plastic bag piece) is proportional to the vacuum created inside the vase.

The weight of the water pushed down the plastic film that bulges (up or down) more near the geometric center of the vessel rim than at its perimeter. The plastic film is a bit adhered to the vase by the surface tension of the layer of water. The sealing between the vase rim and the plastic film caused by the surface tension of the water is fundamental to create a closed vase. However, the adhesion force generated on this sealing and the strength of the plastic film were weak when compared against the pushing down weight that can reach up to 890 gr (1.6 lb.) when the vessel is full of water. This fact is corroborated when too much air enters the vase, equalizing its internal pressure with the external surrounding pressure, vacuum is not created anymore, and all the water is expelled out the vase. Hence, what keeps the water inside of the upside-down vase is mainly the vacuum.

For the next two experiments, find a hard to collapse, transparent, plastic test tube with a lid that closes it hermetically[41], and drill a small orifice near to the bottom of the tube, as the one shown in Figure D26. The first experiment is just a refreshment of an obvious fact. While the orifice is kept plugged by covering it with a finger, fill up water to the brim of tube, then unplug the orifice and let the water flow. The water will start and keep continuously leaking and descending until its top surface reaches the bottom edge of the orifice.

---

[41] Any cylindrical receptacle with a lid like the tin of a typical roll on deodorant can be used instead of the test tube.

On the second experiment, while the orifice is kept plugged, the tube is partially filled with water and then closed hermetically with a lid, so air is kept trapped on its top. Once the orifice is unplugged, the water should start flowing (dropping) just for a while and then stop without reaching the orifice. The water surface shall descend a very small distance that is proportional to the volume of the drained water. During the descending, the water adds volume to the trapped air. This addition of volume to the air stretches its molecules creating a vacuum in the test tube. Thus, the flow of water stops because its draining causes a vacuum, whose effect on the test tube is equivalent to keep a beverage on an intermediate position of the drinking straw by sucking through it just enough.

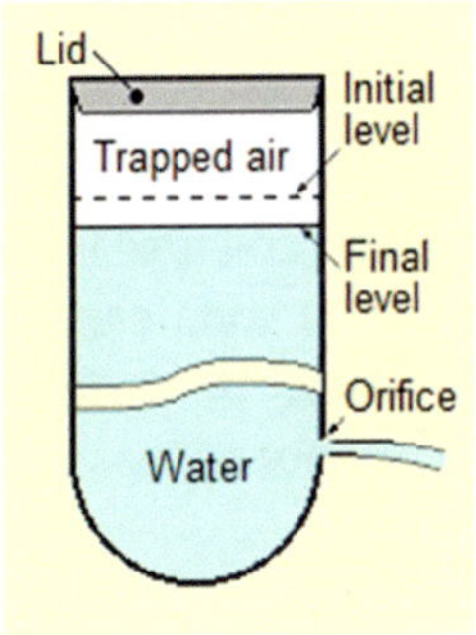

Figure D26. Transparent test tube

To find out the small volume of the drained water or volume added to the trapped air, the following procedure can be used: collect the drained water on a paper napkin, whose dry weight is known, and re-weigh the now partially wet napkin. The difference of weights of the wet and the dry napkin is the weight or mass of the drained water. This mass divided by the water density[42] gives an estimate of its volume. Using this procedure, the estimated

[42] Density of a body is defined as its mass divided by its volume, so the mass divided by its density is equal to the body volume.

volume of the drained water for the second of the experiments was about 0.271 cm^3 (0.0165 in^3). For this volume, the water surface descended about 0.53 mm (0.02 in or nil) in the cylindrical test tube. The tube used had an internal diameter of about 25.4 mm (1 in) that is equivalent to a cross section area of 5.067 cm^2 (0.785 in^2). The volume of a cylinder[43] divided by its cross area is equal to its height, or 0.271/5.067= 0.053 cm = 0.53 mm (0.0165/0.785 = 0.02 in). The volume gained by the trapped air maybe be not equal to the volume of drained water because the air and all gases are susceptible to expand or contract when under the effect of an even moderate pressure. However, on the test tube the created vacuum (negative pressure) is so small, that its effect on expanding the air is nil.

A vacuum cleaner is a small version of an industrial vacuum pump used to extract air, or other gases, from a solid, rigid, and hollow vessel (like a tank) to create vacuum in such equipment. That should not decrease its internal volume under the effect of the vacuum, because it might collapse like a straw does when one suck a beverage trapped in a soda can. A typical test tube is so strong for the vacuum created by draining water that it will not decrease its interior and will not collapse under this small vacuum.

Water, air, vacuum, a balloon, and a canister

Another experiment like the second performed with the test tube, can be (was) conducted with a cylindrical and transparent canister, provided with two orifices, one small near its bottom and another of medium size on the center of its lid. See Figure D27. In case you want to try it, to the perforation on the lid, attach a balloon on such way that it is inside of the canister with its nozzle-mouth opened to the surroundings as shown in Figure D27b. To this purpose, a rubber plug used on wine bottles was modified to hold the balloon and allow its inflation.

---

[43] Volume of a cylinder ($V = \pi r^2 h$) is equal to the area of its cross section (πr^2) times its height (h).

When ready, proceed as follows: partially fill the canister with water while its bottom and lateral small orifice is kept plugged with a finger. Place and tighten the lid. Then, cover the balloon's mouth with a finger, turn upside down the canister and start inflating the balloon while open (unplug) the small orifice to allow the trapped air to escape. Stop the inflation when the surface of the rising water reaches near the small orifice. Keep inflated the balloon, and re-plug[44] the small orifice to avoid the air escapes. Free the balloon from its source of pressure to allow the pressed air inside it to escape through its mouth. You should be good to go.

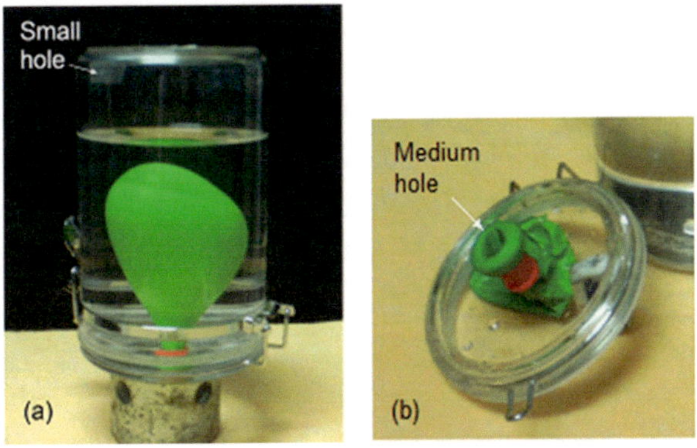

Figure D27. A canister and a balloon

The canister, water, trapped air above the water's surface, and balloon assembly should resemble Figure D27a. The shape of the balloon should be different because it is a function of the uniformness of its elastic membrane and it might vary from one balloon to another.

Once free from the source of pressure, the balloon contracts and squeezes air out until its internal pressure equalizes the surrounding pressure. That should happen in fractions of a second. During the deflation of the balloon, it almost instantly decreases

[44] A piece of scotch tape works for a 1.6 mm (1/16") orifice. For quickness at the beginning it is easy to use a finger to plug the orifice.

its volume, and increases the volume of the trapped air, that creates a vacuum that stops its contraction. What keeps the balloon "inflated" is the sucking of the water and the air in the canister that are under vacuum.

Regarding the experiment described above, prior to inflating the balloon, you can put small objects (bead, pip, a piece of rubber, etc.) on the water and select one that floats, and another that dives, and place them on the canister. You can verify that when they are under vacuum too, they behave the same as they did at the ambient conditions. For instance, the one that sinks down, remains on the bottom under the pressure created by the water under vacuum above it. This fact corroborates that pressure, vacuum, and adhesive forces can coexist in a medium. We in the Earth are also under those three effects or forces, with vacuum being the weaker of all of them.

Pressure and vacuum in a U tube

Positive, and negative pressure or vacuum can be measured by a homemade experiment whose main components are a set of birthday party balloons, a piece of transparent plastic hose or a flexible tube, a wine bottle plug or cork, a sturdy piece of timber, tap water, and a small compressor or air pump. The compressor or air pump is optional. It avoids inflating the balloons using the traditional method, the mouth. Shape the hose into a U having a side or leg shorter than the other, and support it vertically against the timber. Fill the U-tube with water to near the brim of its short side.

When the two sides of the U-tube are open to the surrounding air, they are at the same pressure and their water surface are at the same level, which is labeled as *reference line* in Figure D28. A change between these levels is proportional to the magnitude of the source (pressure or vacuum) that breaks the equilibrium on one of the U-tube sides. The water surface below the reference line is under a higher pressure than the one above it. Attaching an inflated balloon to the short side of the U-shaped hose will create a (positive) pressure that shifts the water surface to different levels on each of the legs of the U-tube. See Figure D28a. The difference between the water surfaces, or height h, is proportional to the remaining pressure in the balloon.

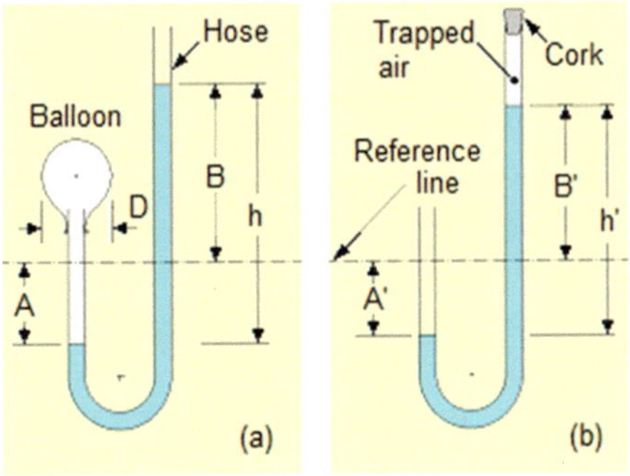

Figure D28. Air under: (a) Pressure, (b) vacuum

Once a column of water or pressure has been created by a balloon, to generate a vacuum install the cork at the end of the long side of the U tube. Then, slowly release the compressed air to avoid that the gust of it reaches the other leg. As soon as the balloon pressure starts decreasing, the water surface on the long side will try to move down to equalize levels with the other surface. This descending action starts stretching out the trapped air, and creates a vacuum on this extreme that stops such action. If the water keeps descending abruptly, or on a steady way, it should be because either the sealing of the plug (cork) is not hermetic or the vacuum is collapsing the hose/tube. The faulty component should be replaced. The created vacuum is proportional to the difference, or height h', between the two water surfaces. See Figure D28b.

If the experiments shown on the sketches of Figure D28 were performed at the same ambient conditions and one just after the other, the corresponding dimensions on the two cases should be almost identical ($A \cong A', B \cong B', etc.$). The differences, if any, are proportional to the descending of the surface of the water caused by the stretching of the air trapped on the long side of the

U-tube. If the internal bore of both U-tube sides have the same diameter, the distances (*A* and *B*, or *A'* and *B'*) above and below the reference line should be the same, and each of them should be a half of its corresponding distance between the two levels of the water surface (heights *h* and *h'*). However, there should be some small differences of few millimeters between these pair of distances caused by small bubbles of air trapped in the water.

Pressure is defined as the ratio of a force to the area of the object that withstands it, and is specified on different units. The most popular are the newton per square meter (N/m^2) known as the *pascal*, the pound per square inch known as *psi*, and inches (millimeters) of water or inches of water gauge known as *in* wg or *column of water*. The late, used commonly for small pressure and vacuum, is called gauge because it is gauged using as reference the atmospheric (barometric) pressure. On this unit, the name of the fluid used is also specified because it could be mercury (or other liquefied substance) instead of water. For a same pressure, the height (*h* or *h'*) varies depending on the fluid used. For instance, mercury has a density greater than the water's, and produces a shorter column than water when both substances are under a pressure or vacuum of the same magnitude.

A membrane body under pressure.

When a balloon is inflated to different sizes, each of these sizes, exerts different pressures over the water surface. Which of them exerts more pressure, a small or a big one? We puff off harder, making more effort at the beginning when a balloon is still small than when it has increased its size. However, by experience we also know that when a balloon is re-inflated, especially, to big sizes several times, it becomes springier and weaker, and it happens because the resilience of its elastic film is not perfect. We detect this weakness by the translucence and how thin its film becomes. On the other hand, the atmospheric or external pressure over a balloon is the same whatever its size. Its internal pressure is associated only to the resistance of its elastic membrane or film.

Thus, the bigger the balloon or bubble, the weaker its membrane and the smaller the internal pressure, or external vacuum, that it can withstand.

Ballon Condition			
New		Re-inflated	
D	h	D	h
125	330	75	372
160	262	103	298
225	215	155	242

Note: D and h (pressure) in milimeters (a)

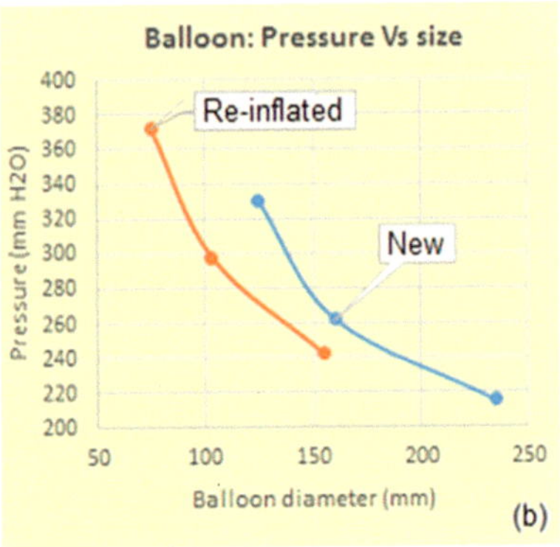

Figure D29. Pressure exerted by a balloon

These facts are confirmed by the information shown in Figure D29 that was obtained for a balloon using the U-tube of the previous section. It is worthy to notice that the diameter "D" is estimated because a balloon gets several round bulged shapes but, with few exceptions, spherical. On a group of concentric spherical membranes or bubbles which ones have high inside pressure, the more internal of the external ones? According to the experiments described on previous paragraphs and to our knowledge about the strata of our planet, the internal is under high pressure.

A ruler, an invisible wheel, and an old curve

This section includes two experiments, the first one is as a physical reminder of the concept of a circle: a set of points in a plane, like the rim of a wheel, that are at the same distance from other point known as center, like the wheel axle. The second experiment reveals that circles might be and are embedded into certain bodies and/or associated to them. It consists of just bending a flexible and flat object having elongated shape like a regular ruler.

For the first experiment the following materials are needed: a small object that weighs a few grams like a metallic washer or nut, a piece of thin twine or thread, a small disk of wood or a circular plastic lip of about 100 mm (four inches), and two straight pins to be part of the disk, one as its axle or shaft and the other to be pierced near to its perimeter. The idle wheel of an upside-down bicycle can be used instead of the disk that might be transformed into a "wheel" by the axle pin.

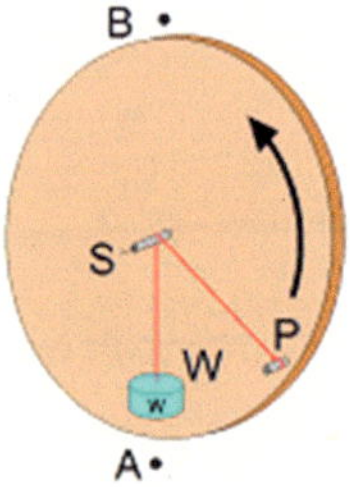

Figure D30. Weight hanging from a wheel

As shown in Figure D30, to set the components, tie to the peripheral pin one end of the twine, and on its other end attach the weight, then slowly turn the disk until the peripheral pin reaches the bottom and the weight is hanging from the axle. Be sure the twine is in contact with the axle and is not wound around it. Outside and near to the periphery of the disk, mark and identify two spots, one on the bottom (A) and the other (B) diametrically opposed to the first on the top.

To carry on the experiment, interchange the position of the peripheral pin by slowly turning the disk (wheel). Be sure that the length section of the twine in contact with the axle starts unwinding around it. By doing so, the hanging weight *w* will remain static on a fixed position. It happens because the twine length from the pin *p* to the spinning axis is equal to the radius of the circle described by the spinning pin *P*. Depending on the diameter of the axle and/or the twine, the weight may move slightly.

However, if the axle diameter and twine thickness were as small as a human hair or less, the twine spot in contact with the axle, will be as close as possible to the imaginary spinning axis of the disk (wheel), and the weight will not move at all. For the arrangement shown, the weight will remain static for half revolution of the wheel. After more than one revolution in the same direction, the twine will move to the other (right) side of the axle and start winding around the spinning axis and the weight will ascend.

On the next and the second experiment we are going to discover that when a flexible ruler or straightedge is bent it adopts a bending profile that is an old mathematical (hyperbolic spiral[45]) curve. That happens to be the same way as decay the rotational speed of spinning bodies just under the effect of friction. Many of the bodies used on the experiments of previous parts behave in this way. This experiment can be done and was done with a flexible ruler (one on stainless steel with an overall length of 615 mm-24.2 inches), a big and dull needle to be attached at the ruler midpoint, a metallic washer, a piece of twine, a small metallic nail, two small C-clamps, a tape measure, and a timber of about 64 mm (2.5 inches) width, 20 mm (0.75 inches) or more thick and with a length of about 1160 mm (42 inches).

To set the experiment, cut the timber into three pieces: one to be used as straightedge 230 mm (nine inches) longer than the ruler, and two small pieces of about 100 mm (four inches) long each. Place the longest piece on a working table[46], and using a clamp attach to one of its ends one of the short pieces as it is shown in Figure D31. On the lateral side of the straightedge, place the nail about 9.5 mm (3/8 in) from and below its top edge, and about 45 mm (1¾ in) from the edge of the attached piece of

---

[45] This curve was studied in detail by Pierre Varignon (1654-1722).

[46] The straightedge may not be necessary if the table is long enough.

timber. This location of the nail was obtained through trial and error and could vary for a ruler of different material and/or dimensions. Over the straightedge, place the ruler so that it is in contact with the attached piece of wood that shall act as a pivot for an end of the ruler. Tie one end of the twine to the needle eye, and the washer to the other end. Pass the free end of the twine around and over the top of the nail on such way that the hanging washer is free to move up and down.

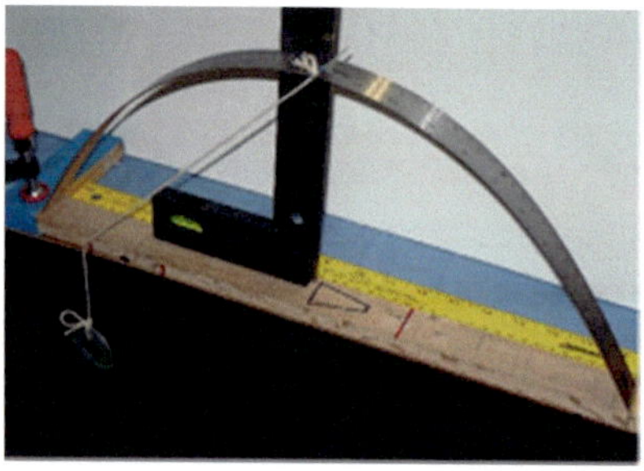

Figure D31. A bent flexible ruler

To perform the experiment, use the another small piece of wood to slowly push the free end, the one far from the nail, toward its pivoted end or the one near to the nail. The ruler shall bend and haul the needle while its free end is moved over the straightedge following a straight line. The picture of Figure D31 shows an intermediate position where the needle has moved up about 174 mm (6-7/8") and about 76 mm (3") toward the pivoted end. When the ruler was on its flat condition and resting on a wood straightedge, the needle was over the red line mark that is located about 305 mm (12 inches) of one end of the ruler. To keep the flexible straightedge bent on one position, fix the pushing piece of wood using the second C-clamp (these components are not shown in the picture).

On the experiment performed, in order to avoid a permanent (not elastic) deformation due to excessive bending, the ruler ends were approached distances a bit greater or less than the distance going up by the ruler midpoint. An aluminum ruler having the same length and width as the stainless-steel one but about 2 mm (5/64 in) thicker, got a permanent bend when the bent reached a proportion or ratio much smaller than the described above. In contrast, a slender steel strip bent until its ends met, and it got an elongated omega (Ω) shape without a permanent (plastic) deformation.

During the ends approaching process and with the nail at the location shown in Figure D31, the ruler bent and the needle moved up and toward the pivoted end, but the washer remained at one fixed position mimicking the behavior of the weight hanging on the spinning wheel described previously and sketched in Figure D30. It seems there is an invisible wheel whose spinning axis coincides with the nail location, through which the twine passes around; it also seems that the needle takes the place of pin P. Thus, the needle followed an invisible circle whose radius is the distance between the needle and the nail. Why does the ruler midpoint behave in such a way?

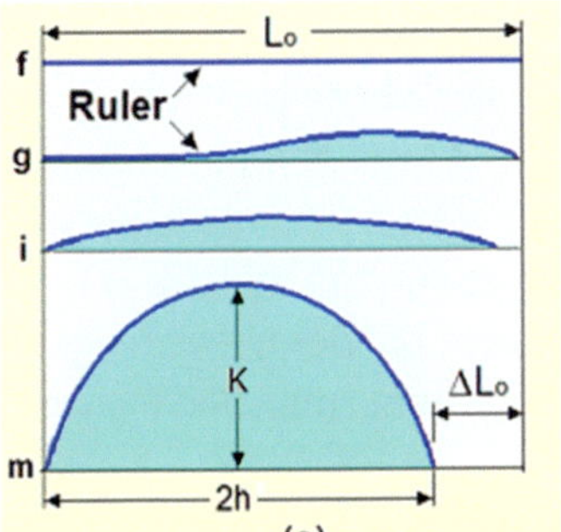

(a)

Curve	2h	K	ΔL_0	$\Delta L_0/Lo$	L'
g	---	16	1	0.16	---
i	613	24	2	0.33	615.5
j	610	38	5	0.81	616.3
m	550	123	65	10.57	616.4
n	460	180	155	25.20	608.0
o	390	208	225	36.59	596.7

Distance in milimeters

(b)

Figure D32. A bent ruler: (a) profiles, (b) dimensions

During the experiment almost all the profiles of the bent of a solid flexible strip resembled parabola curves (their profiles resembling those of the ends of a chicken egg). Were the profiles actually parabolas? To answer these questions, let's detail a bit of data (units in mm) obtained from the experiment. Figure D32a shows several of these profiles identified by a letter at their left side, and the key (k, 2h, and ΔL_o) dimensions that differentiate and "quantify" a ruler bending. The top straight and thick (f) line represents the flat and straight ruler whose length is named L_o. The bottom (m) sketch shows the main three dimensions.

Regarding sketch labelled "g," notice that the bulged section of the curve is not at the flexible straightedge midpoint; it happened because the approached length (ΔL_o) was so small (about 1 mm or about 0.16% of the ruler length). Several attempts were made to promote a bulge at the midpoint, but the bulge always walked away from the midpoint and toward one of the ruler ends. At this small ΔL_o value, it seems that the whole flexible straightedge follows the Principle of Least Action. First stated by Pierre-Louis Moreau de Maupertuis (1698-1759) as "any changes in nature are such as to make the action least" (Kline 581). This principle was also stated by Leonard Euler (1707-1783) as "all natural phenomena behave so as to maximize or minimize some function" (Kline 582).

In order to compare the bent ruler profiles against parabola curves, and knowing that all the profiles of the bent straightedge have a fixed (L_O) length of 615 mm, it was compared against a parabola length. That was calculated with a formula expressed in function of two parameters, specifically the height (K) at its midpoint and (h) the half of the distance between (opening) ends. The same parameters measured on the bent ruler. See Figure D31, and sketch *m* of Figure D32a.

The length values obtained are shown on column *L'* of table of Figure D32b. The results of the mathematical formula are not accurate because the input parameters (h and K) were measured with a tape measure. For small values of about 10% or less of the ruler ends approaching (ΔL_o) distance, the values on the calculated length almost match with the actual length of the flexible straightedge. For small bends it is possible to infer that the shapes obtained by bending a ruler are close to the shape of a parabola.

However, for high bending, these bends are not following parabola curves.

For comparison purposes, profiles of a flexible rectangular ruler were first traced on a tracing paper from actual bents of the ruler, and then copied over a set of hyperbolic spiral curves (for details about this curve see *The circle and the hyperbolic spiral curve)*. For one particular curve, this process produced a sketch similar to the one shown in Figure D33 for three (*H, L, M)* profiles (*M* not shown for clarity). Where key information of these profiles is listed on attached table *II*.

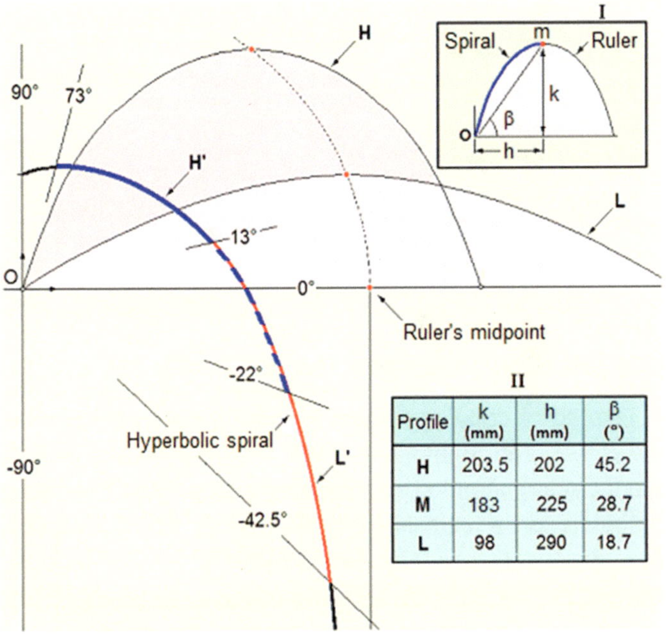

Profile	k (mm)	h (mm)	β (°)
H	203.5	202	45.2
M	183	225	28.7
L	98	290	18.7

Figure D33. A ruler and the hyperbolic spiral curve

From this sketch, it was discovered that the half branch of each ruler profile (from its *O* end intersecting with the approaching line to its m midpoint, *see Sketch I*) matches with a segment of a hyperbolic spiral curve. For instance, the half of the *H* profile matches with the segment H' (top sector on solid and dashed blue line) of the hyperbolic spiral going from about 73 degrees to minus 22 degrees; the half of *L* profile matches with the segment *L'* (bottom sector starting on dashed red line) going from an angle of about 12.5 degrees to minus 42.5 degrees; and the corresponding half of the *M* profile matches with the sector going from about 61 degrees to minus 26 degrees. So, when the ends of a flexible rectangular ruler approach to each other following a straight line, the ruler (Figure D31) bends into profiles that are segments of a hyperbolic spiral.

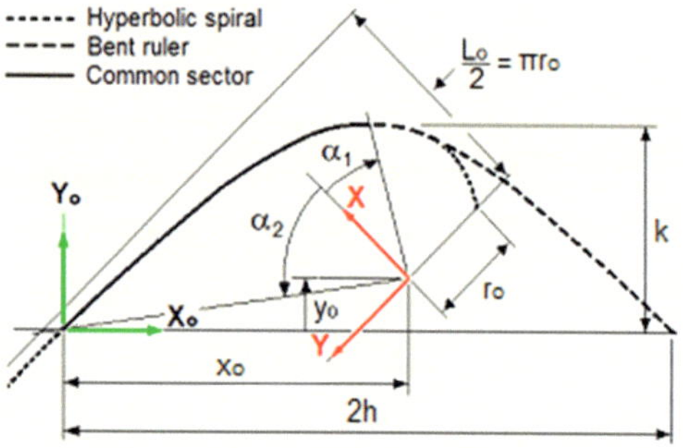

Figure D34. Hyperbolic spiral curve and profiles of a bent ruler.

Instead of copying a ruler bending profile with the help of tracing paper, and manually make it match with a segment of a hyperbolic spiral; should this matching process be done using just mathematics or a calculation? This is an open question for mathematicians. Figure D34 could be helpful for this purpose. On this sketch, the X_oY_o coordinates axes of the straightedge is hinged-fixed at the meeting point while the XY axes of the hyperbolic spiral will change location as the ruler bends while its ends are approaching.

It is amazing that the exotic hyperbolic spiral curve is on a hidden way related to circles, or that one straight movement is converted to a circular one! It seems that everything on this spherical world is in some way associated to circles. Are we obeying a cycling law? Are we cycling the same as our planet does around the Sun, or the way electrons do around the atom nucleus?

Part E Determining properties of solids

It isn't that they can't see the solution. It is that they can't see the problem.
G.K. Chesterton

For having internal voids or small spots of different density, the bodies usually are not uniform and their center of gravity normally does not match with their geometrical center. On rotating bodies having a symmetrical and/or round cross section, the purpose of balancing them is to make coincide these two centers. To get them balanced, weight (mass) is added or removed from the body, and to determine where to do that calculations have to be done. On the next section is detailed one calculation method. Adding weight is avoided on some occasions because: 1) the attached weight could be a protuberance that would cause an unbalance on the spinning plane; 2) the attaching of weight process could be a time consuming task; 3) this process could create abnormalities on the material of the body; 4) the added material can be expelled away by the centrifugal force creating an unexpected unbalance that even on a short period of time could generate a catastrophic effect on the performance of a component spinning at high speed. For example, an unbalanced turbine spinning at several thousands of rpm could generate destructive vibrations that not only will destruct it in a few minutes, but also the traveling airplane to which it is attached. However, when none of these reasons is a big issue, the adding weight is preferred over the removing weight. For instance, a car wheel is balanced snapping weight to its rim, because if the added weight is thrown away, the vibration has no destructive effects due to the relative low spinning speed of the wheel.

When some mass is lost from or added to a body, it immediately adjusts the location of its center of gravity. The body is not waiting for any calculation to know the new position of the center of gravity. If the change of mass (weight) happens on a rotating body, the adjustment is also done immediately, and an unbalance will be evident if there is enough spinning speed to promote vibration. It seems energy (nature) is looking for a way to reverse the mass change or a way to minimize its effect. If a small body were glued to the perimeter of a rotating wheel of uniform density, the wheel could/might vibrate until the external body is expelled away from its perimeter. If there is not enough spinning, could the wheel spread out the strange body over its perimeter? If the spreading were uniform over the entire perimeter, the strange body would not change the center of mass of the wheel. A similar spreading effect seems to occur on the face of Earth by the tides of the oceans.

The next section will present a method to calculate the center of gravity of a body when it losses or gains some mass.

Balancing and the Add-cut theorem

The old and known seesaw seen on must public kid's playgrounds, is a plank or bar that can turn around its centered support known as fulcrum. It was conceived and stated on some Archimedes (287-212 B.C.) postulates and propositions about the lever and center of gravity. Imagine a small object of mass *m* tied to a string and hanging from a balloon full of the lifting helium, and when the small body-balloon couple is released into the air, the couple remains floating at one level without going up or down as shown in Figure E1c. On this balanced condition, the balloon exerts an upwards force of magnitude equal to the downwards object weight.

Also imagine, that on each of two long and big identical objects of mass *M***,** their center of gravity have been identified and marked as point *o* and a distance *L* from this point another has been marked as *p*. Each of these bodies is placed over two hypothetical weightless levers. Each of these levers is supported by a fulcrum which is at a distance *k* from the center of gravity of the big body just put on it. After separating the couple by cutting their interconnecting string, the small object is located on one of the big bodies over its point *p* as shown in Figure E1a, and the balloon is tied to the other point *p* of the other body as illustrated in Figure E1b. On these arrangements, what should be the location (right or left) of the fulcrum and the distance *k* from the center of gravity (point o) of each of the two big objects to get them balanced? To answer these questions, let's use two well-known Archimedes' propositions.

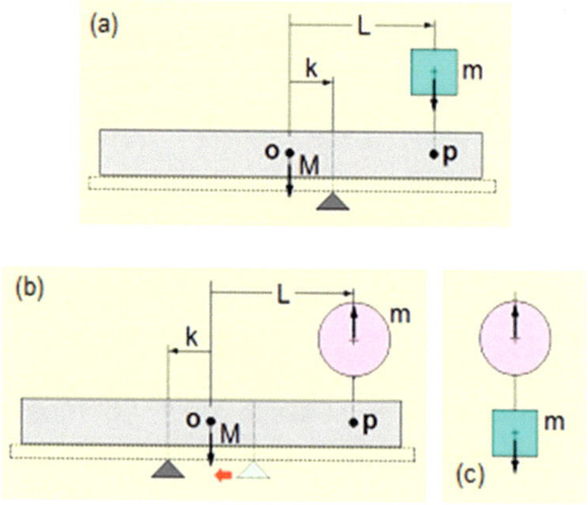

Figure E1. Long bars on planks supported on a fulcrum

The first of these propositions is stated as: "Two magnitudes whether commensurable or incommensurable balance at distances reciprocally proportional to the magnitudes" (Kline 164). In this statement, reciprocally proportional means acting on opposite direction, and when these two magnitude-distance and opposite effects (torques) are equal, they balance. Noticing in Figure E1a that the distance *k* times mass *M* about the fulcrum is opposed in turning direction to mass *m* times the $(L - k)$ distance from the fulcrum to point *p*. These products acting on opposite direction should be equal to be balanced and avoid the lever turning about the fulcrum. In mathematical terms is

$$M.k = m.(L - k),$$

That after arranging and leaving *k* alone becomes

$$k = \frac{m}{(M+m)}L \qquad \text{(E.1)}$$

If the arrangement of forces in Figure E1b were the same as the one in Figure E1a, the mass-distance couple on the balloon for its up effect would make both couples to turn on the same direction. Consequently, the fulcrum should be moved to the left of point *o* to make the couples aim on opposite direction. Making this change both (o and p) points should be at the right of the fulcrum, and the distance from the fulcrum to point *p* is *k* + *L*, so the balance will occur when

$$M.k = m.(L+k)\,,$$

that after arranging and leaving *k* alone becomes

$$k = \frac{m}{(M-m)}L \qquad \text{(E.2)}$$

For being weightless, the lever (plank) does not affect the results gotten by the above equations, so the relocation of the fulcrum is equivalent to moving the center of gravity of the big body. In other words, the location of the fulcrum must coincide with the center of gravity to keep the big body balanced. This fact is used as a practical way to find the center of gravity on the longitudinal axis of elongated bodies. For example, on a boat by balancing it over a log used as the balancing point or fulcrum. We know that the boat center of gravity is at or near the spot where the log is contacting the boat, when none of its ends are touching the ground (very unlikely that happens), or when we can easily lift the boat end off the ground. The easier the lift, the closer the center of gravity to the log.

The mass-distance couple, or the effect of a force promoting a turning action around a distant point or axis, is known as *moment of a force* (mass). A formal way to apply the propositions and postulates of Archimedes stated above is imagining the fulcrum as the shaft of the arms of a wall clock, around which should be moving on opposite directions the mass-distance couples.

The lifting action of the balloon is equivalent to making lighter the side of the big bar where the balloon is located, for instance, extracting mass from such side and spot of the bar by drilling a hole. From the discussion on previous paragraphs, it is possible to infer that the new balancing (fulcrum) point needs to be moved from the previous center of gravity towards the added weight, or away from the removed weight.

Keeping in mind the facts described on above paragraphs, let's use another proposition of Archimedes. Namely "if two weights have not the same center of gravity, the center of gravity of both taken together is at the middle point of the *line joining their center of gravity*" (Kline 164). Let's imagine that for a flat body of any shape, uniform density, and unique thickness, the location of its center of gravity, or center of mass is known. This center could be obtained by a trial and error process for bodies amorphous or complex in shape as it was suggested above on balancing a boat. For symmetrical bodies it could be found by mathematics, and in some cases just by inspection.

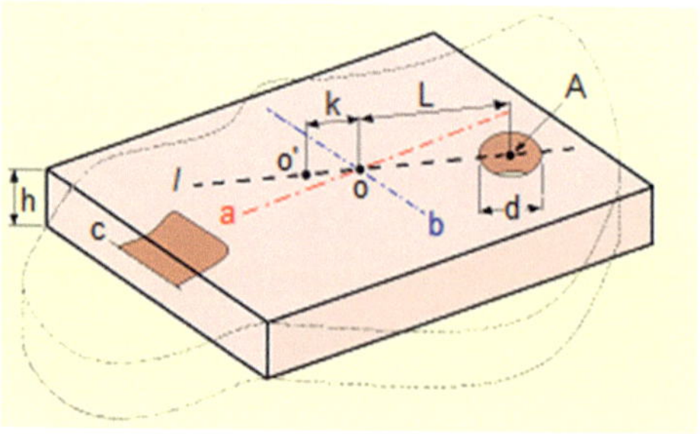

Figure E2. Center of mass of a flat body

For a rectangular solid body like the one shown in Figure E2, this center is at the midpoint of both width and length of the body, or over the intersection of its *a* and *b* lines of symmetry. If the initial mass (*M*) of either of the objects of Figure E2 is known, and a mass (m) is cut out of it by drilling a hole at point *A*, that is at a distance (*L*) from the actual and known center of mass (*O*) of the body. For Archimedes' postulate, its new center of mass (*O'*) must be located on the straight line (*l1*) crossing through point A and point *O*. If a snapshot of the cross section of the body is taken over this line, it will resemble Figure E1b where the fulcrum is already located on the new center *O'* that is at a distance *k* from the initial center of mass *O*. Due to this resemblance, this *k* value is given by Equation E.2.

The procedure described above suggests that when we have two bodies whose center of mass, and mass are known, we can obtain a third one by adding or subtracting one of them to the other. Using this procedure of adding, moving, or cutting one object to/from other; let's find the center of gravity of a hypothetical can that is composed mainly by a hollow cylinder having a flat disk attached in its inside to one of its ends as shown on the left sketch of Figure E3. By symmetry and observation, its center of gravity is on the imaginary and symmetrical line *a* going through the middle of the cylinder, and should be moved toward the disk that makes the body asymmetric regarding the length of the body.

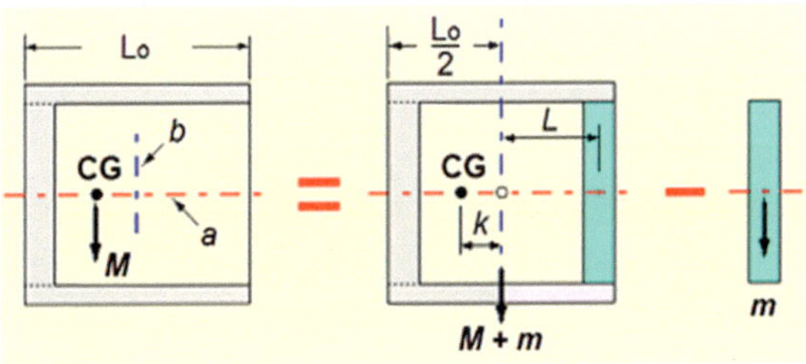

Figure E3. Can composed of a cylinder and a disk

The center of gravity (CG) of this can is not at the midpoint of the cylinder height. When a fictitious disk identical to the actual one is added to the open end of the can, the transformed body is now also symmetrical to the length of the cylinder for having a disk on each of its ends. The center of gravity of this body is at the middle of the cylinder length (line *b*) or at Lo/2. This arrangement resembles Figure E1b, where the transformed would stand for the big object, and the fictitious disk would stand for the balloon, or the small object to be removed or cut from the big body. The center of gravity of the actual can, or the distance *k*, can be found using Equation E.2.

Between the Earth and the moon there is a point known as the barycenter, or common center of mass around which the Earth and Moon are orbiting each other. The position of this point can be found using Equation E.1. In this case, the average Earth-moon distance (L) is 3.84×10^{8} m, the Earth mass (M) is 5.98×10^{24} kg, the Moon mass (m) is 7.36×10^{22} kg, and k is the distance from the geometric center of the Earth to this point. Its value is:

$$k = \frac{m}{M+m} L = \frac{7.36\ x\ 10^{22}}{5.98\ x\ 10^{24} + 7.36\ x\ 10^{22}} 3.84\ x\ 10^{8}$$

$$k = 4.67x10^{6}\ m$$

As the average radius of the Earth is 6.37×10^{6} m, the average barycenter is inside the Earth at about three fourth ($4.67/6.37 \approx 0.73$) of its radius. Even if the mass of the Earth-Moon system is constant, its center of mass (barycenter) is on a continuous change because the (L) distance between them varies during the orbiting of the moon.

Adding and/or cutting out mass (weight) to get the location of the center of gravity of a body is the same technique used to balance a car wheel and other machine-rotating components. For that reason, the procedure to get either Equation E.1 or Equation E.2 can be also labeled as *the Add-cut theorem*.

Center of mass of a split key ring

The center of gravity (fictitious or actual point in which is concentrated the mass of a body), and moment of inertia and other properties of common of bodies are normally calculated assuming homogeneous and symmetric objects. Such calculations for asymmetrical and not homogeneous objects like a split key ring will be introduced in this and the next section.

A differential is a relative concept that is fundamental in advanced mathematics, and all of us, even if we are not a mathematician, at least have a practical knowledge of it. For instance, if we want to place a fence around a lot or pool having an amorphous figure, we surround it with vertical support poles placed a short distance among them. Although the wood crosspieces and/or wire lines between two adjacent poles usually are on straight line, the fence adopts the shape of the contour of the pool. The distance between two adjacent poles is what in mathematics is called a tiny difference, or a differential of length. In this example, when all of them are added up we get either very close to, or the exact value of the total length or perimeter of the pool. It could also be even much bigger and still be a differential. For instance if you are in the middle of a vast prairie, there is a big chance that the land two or three miles around from your position appears flat to you. That vast flat land around you could be a differential of area of the Earth. That is, or appear to be, a sphere when one watches it from the Hubble space telescope which is about 560 km (350 mi) from Earth surface.

A differential which are used among other things to get some physics properties of an object, is denoted by the letter *d* followed by another letter representing the variable in consideration. For instance, *d*L is read as differential of length when *L* stands for length. It is worth to note that a differential is a kind of close up of a trait of an object. It should be small enough to appear to be flat and/or straight and be the essence of the topic of consideration. That is, when the differentials are summed up their total shall be very close, or equal to the actual total trait (area, volume, etc.) of the object. Figure E4 shows the mass and the length of very tiny, infinitesimal (d_m and d_L), or differential sections of a solid of uniform density.

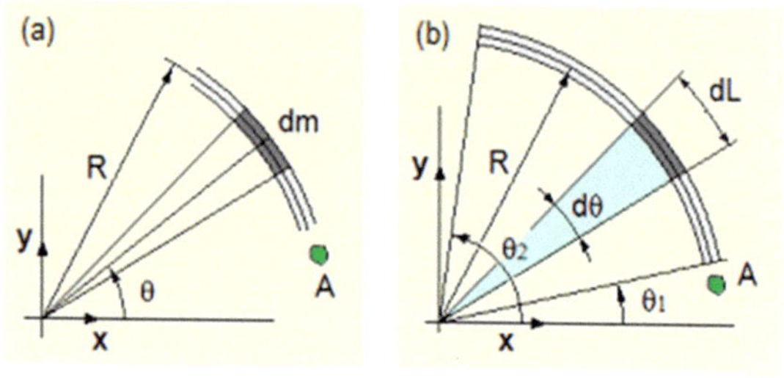

Figure E4. Tiny sections of mass and length of a ring

On an infinitesimal section of a ring of uniform density ρ, cross section of unique area *A*, and length *dL* (see Figure E4) equal to $dL = Rd\theta$, its volume is $dv = AdL$ or $dv = ARd\theta$. On the other hand, the mass of a solid object of uniform density (ρ), is proportional to its volume, and is defined as mass equal to density times volume, or *Mass = ρ x Volume*. A differential of this mass is $dm = \rho dv$ that after using the value defined previously becomes

$$dm = \rho ARd\theta \qquad \text{(E.3)}$$

If instead of differential of mass *dm*, we have a small sector of the ring of mass *m* and limited by angle ψ, the above equation becomes

$$m = \rho AR\psi \text{..............(E.3a)}$$

When a solid figure of uniform density is flat and straight like a timber, or just straight like a bar, its center of mass is inside the body and located in the middle of its three primary dimensions, but any kind of bending could move this center even outside of the body. A typical example is a solid ring of uniform density and unique cross section in which the center of mass does not touch the ring and coincides with its geometric center.

For a section of a solid ring of radius R, and total mass M_S, the location of its center of mass (*CG*) is defined (Serway and Jewett 246) as, see Figure E5a:

$$\bar{x} = \frac{1}{M_s}\int r_x dm \ ; \ \bar{y} = \frac{1}{M_s}\int r_y \, dm \ ; \ \bar{R} = \sqrt{\bar{x}^2 + \bar{y}^2}$$

Where $\bar{x}$ and $\bar{y}$ are the coordinates of *CG*, r_x and r_y are the coordinates of one of the multiple tiny portions of mass dm forming the section. The enlarged s-shape symbol is known as *integral* and it represents the sum of all the items at is right or inside it. The solutions of typical and common integrals to be presented in further paragraph are get from a table of integrals of a calculus book.

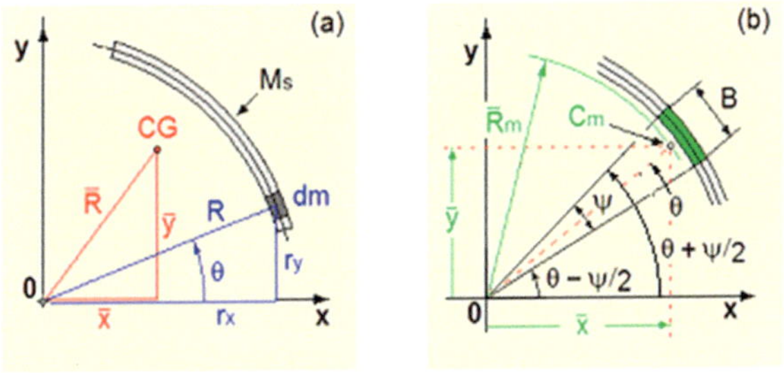

Figure E5. Center of mass of ring sectors

For a section of ring limited between θ_1 and θ_2 angles, see Figure E4b, and having a differential of mass given by Equation E.3; which r_y component is given by $r_y = R\ sin\theta$ (Figure E5a). The ordinate of its center of mass is given by:

$$\bar{y} = \frac{1}{M_s}\int (R\ sin\theta).(\rho A R d\theta) = \frac{\rho A R^2}{M_s}\int_{\theta_1}^{\theta_2} sin\theta\ d\theta$$

$$\overline{y} = \frac{\rho A R^2}{M_s}(cos\theta_1 - cos\theta_2) \tag{E.4}$$

Similarly, for r_x component given by $r_x = R\cos\theta$, the abscissa value is

$$\overline{x} = \frac{\rho A R^2}{M_s}(\sin\theta_2 - \sin\theta_1) \tag{E.5}$$

The radius of the center of mass of this section, regarding the geometric center of the ring from which the ring's section was extracted, and from equation E.4 and equation E.5, is obtained by

$$\overline{R} = \sqrt{\overline{x}^2 + \overline{y}^2} = \frac{\rho A R^2}{M_s}\sqrt{2 - 2\cos(\theta_2 - \theta_1)}$$

According to this equation, for a complete ring where $\theta_1 = 0$ and $\theta_2 = 360$ degrees (2π), $\overline{R}$ is zero. That is, the center of mass is in the geometric center of the ring and not touching it. That is a well-known fact. In the above relationship and in some cases, the trigonometry double-angle identity ($\text{Cos}2\theta = 1 - 2\sin^2\theta$) could be very useful.

Using the above equation for a small sector of a ring, like the mouth of a split key ring, limited for an angle ψ between $\theta 1 = \theta - \psi/2$ and $\theta 2 = \theta + \psi/2$, (see Figure E5b), which M_s mass is given by equation E.3a, this radius is obtained by

$$\overline{R}_m = \frac{2\rho A R^2}{\rho A R\psi}\sin\psi/2$$

$$\bar{R}_m = R\frac{sin\,\psi/2}{\psi/2}$$

The ratio $(\sin \psi/2)/(\psi/2)$ is almost one for small angles ψ. For the split key ring used in other sections *this* angle is about 13.5 degrees, and this ratio is 0.99077, so $\overline{R}_m \simeq R$. Thus, this relationship suggests things that we know by intuition for a very short piece of a curved stick, the center of mass is not dependent on the angle θ and shall be close to the body.

Using *The Add-cut theorem*, it is possible to find the distance from the geometric center to center of mass, or eccentricity *e*, of a split key ring. This ring is made of two parallel rings, and one of them having a mouth of width *B* and mass *m*. If the split key ring were made of two complete rings, its center of mass should coincide with its geometry center ($\overline{R}_m = 0$). According to equation E.3a and for θ angle of 2π, the mass of just a ring is $\rho AR(2\pi)$. Therefore, the mass of the two rings is twice such value or

$$M = 4\pi\rho AR.$$

As the mouth was formed by cutting out or extracting it from one of the rings, the applicable relationship is Equation E.2. In this equation, the distance *L* from the geometric center to center of mass of the mouth (C_m in Figure E6), is equal to $\overline{R}_m$, and *k* is the eccentricity *e*. Hence, its value is

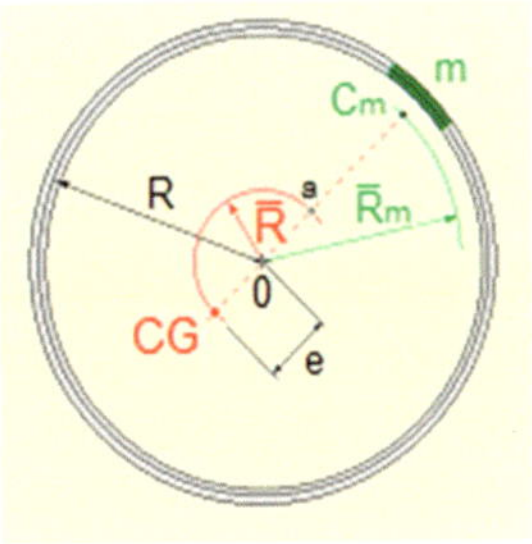

Figure E6. Eccentricity of a split ring

$$e = \frac{m}{(M-m)} L = \frac{\rho AR\psi}{4\pi\rho AR - \rho AR\psi}(R\frac{\sin\psi/2}{\psi/2})$$

$$e = \frac{2R\sin{}^{\psi}/_{2}}{4\pi - \psi}$$

Assuming the mouth section is straight and from half of the triangle its forms with angle ψ (see Figure E5b), $B \cong 2R\sin{}^{\Psi}/_{2}$, thus

$$e = \frac{B}{4\pi - \psi} \qquad \text{(E.6)}$$

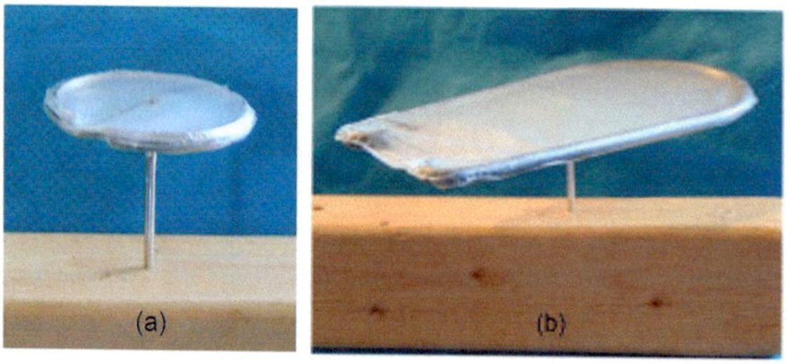

Figure E7. Bodies in equilibrium

To validate Equation E.6, a split key ring whose center of mass (*CG*) is away a distance *e* from the geometric center, was wrapped with a layer of thin sticking paper. Then, such center was painted/drew over the line passing through both the ring's geometric center and the middle of the width (*B*) of the mouth. When the wrapped ring was placed on this spot (*CG*) over the dull end of a vertical needle, it got equilibrium at an almost perfect horizontal level as it is shown in Figure E7a. A similar experiment was done also for a double-hook (Figure E7b).

Moment of inertia of some round bodies

In general, the moment of inertia (I) regarding an axis in a plane crossing a body, is defined as the body mass times the square of the distance from the axis to all the spots and all cross sections shaping the body. As a body can be crossed by multiple planes each generating a cross section, whose boundary can be approached from different directions, the moment of inertia of a body is not unique. It is a kind of property that measure the tendency of a moving body to keep moving and/or stay in its line, or plane of movement.

For a ring spinning about y axis, the moment of inertia of a very small (differential) section of the ring of mass dm whose center of mass is a distance *r* from the y axis (see Figure E8a) in math terms is defined as (Serway and Jewett 276):

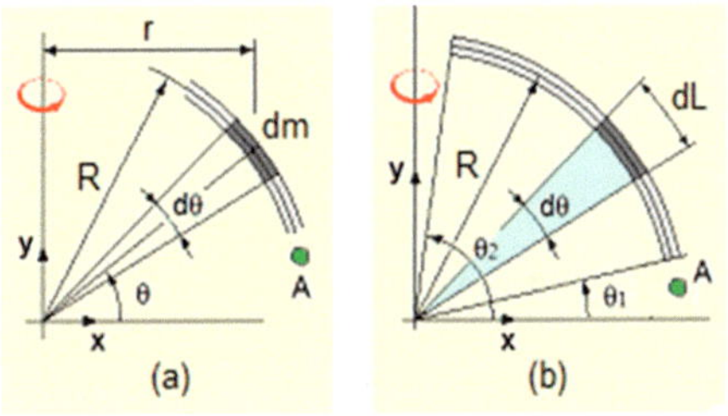

Figure E8. Moment of inertia of sections of a ring

$$I_y = \int (r)^2\, dm \qquad \text{(E.7)}$$

Where

$$r = R\cos\theta \quad \rightarrow$$

$$(r)^2 = R^2\cos^2\theta \qquad \text{(E.8)}$$

Placing equation E.8 into equation E.7, and using the differential of mass given by Equation E.3, we can express the moment of inertia for a section of ring of cross section *A* and uniform density ρ, and limited between θ_1 and θ_2 angles, as the one shown in Figure E8b, by

$$I_y = \int_{\theta_1}^{\theta_2} (R^2 cos^2\theta)(\rho A R d\theta) = \rho A R^3 \int_{\theta_1}^{\theta_2} cos^2\theta d\theta$$

The solution of this integral is

$$I_y = \rho A R^3 \left[\tfrac{1}{2}\theta + \tfrac{1}{4} sin2\theta\right]_{\theta_1}^{\theta_2}$$

, or

$$I_y = \tfrac{1}{2}\rho A R^3\left[(\theta_2 - \theta_1) + \tfrac{1}{2}(sin2\theta_2 - sin2\theta_1)\right] \quad \text{(E.9)}$$

For a quarter of a ring where $\theta_1 = 0$ and $\theta_2 = \pi/2$, see Figure E8, equation E.9 becomes:

$$I_{y(\frac{1}{4})} = \frac{\pi}{4}\rho A R^3$$

For a whole solid ring having a total mass equal to $M = 2\pi\rho AR$, its moment of inertia about a vertical axis passing through its geometric center is four times $I_{y(\frac{1}{4})}$

$$I_y = 4\left(\frac{\pi}{4}\rho A R^3\right) = \pi\rho A R^3$$

$$I_y = \tfrac{1}{2} M R^2 \quad \text{(E.10)}$$

For a solid ring (see Figure A11b), its moment of inertia about the rolling (r) axis which is circumscribed by the ring's mass, r goes across its body without touching it and passes through its center of mass, is

$$I_r = MR^2 \quad \text{(E.11)}$$

The term on the right side of the above relationship happens to be the same for a ring and a disk both having the same radius and thickness. However, because a disk has more mass than a ring, its moment of inertia is also bigger.

The moment of inertia of a sphere regarding an axis passing through its geometry center can be found by a procedure similar to the one described on previous paragraphs, and is given by (Serway and Jewett 278)

$$I_{c(sphere)} = \tfrac{2}{5}MR^2 \qquad \text{(E.12)}$$

Where M is the sphere mass and R its radius.

Assuming that the split key ring mouth, section *m* in Figure E9, covers an angle ψ of the ring, and it has the same transversal section area *A*, and uniform density ρ as the split key ring itself. According to Equation E.9 for $\theta_1 = \theta - \psi/2$ and $\theta_2 = \theta + \psi/2$, and for the product-to-sum trigonometry identity $2\cos a \sin b = \sin(a+b) - \sin(a-b)$, the moment of inertia of this mouth, is given by

$$I_{y(m)} = \tfrac{1}{2}\rho AR^3(\psi + Cos\, 2\theta\, Sin\, \psi) \qquad \text{(E.13)}$$

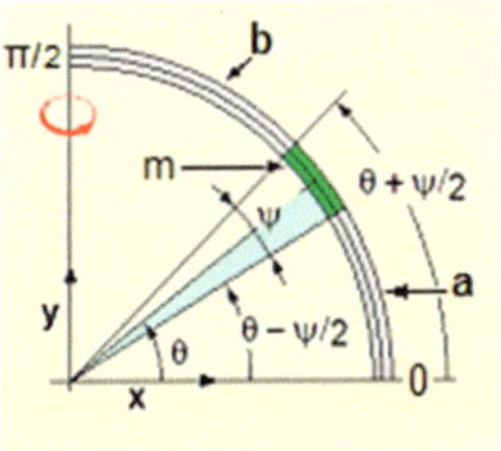

Figure E9. A quarter of split ring and the mouth

According to Equation E.13, the moment of inertia of the mouth about y axis is directly proportional to its position on the ring, that is, a function of the angle θ. When this angle (θ) is either 0 or 180 degrees, the moment of inertia is maximum. That is in accordance with what a figure skater does to suddenly stop the spinning[47], he or she gets his or her legs and arms away from the vertical spinning axis to increase the moment of inertia.

Also for equation E.9 for the *(a)* section of the ring limited by angles $\theta_1 = 0$ and $\theta_2 = \theta - \psi/2$, see Figure E9, the moment of inertia is:

$$I_{y(a)} = \frac{1}{2}\rho A R^3 \left[\theta - \frac{\psi}{2} + \frac{1}{2}\sin\left(2\theta - \psi\right)\right]$$

Similarly for the (b) section limited by angles $\theta_1 = \theta + \psi/2$ and $\theta_2 = \pi/2$, the moment of inertia is:

$$I_{y(b)} = \frac{1}{2}\rho A R^3 \left[\frac{\pi}{2} - \theta - \frac{\psi}{2} - \frac{1}{2}\sin\left(2\theta + \psi\right)\right]$$

From the above equations, for a quarter of a circle and any small angle ψ at any angle θ, see Figure E9, it is obtained that

$$I_{y(\frac{1}{4})} = I_{y(a)} + I_{y(b)} + I_{y(m)}$$

Therefore, the moment of inertia obeys the superposition principle that can be applied to get the moment of inertia of a quasi-symmetrical solid body like the split key ring.

By the superposition principle, for the split key ring turning about *y* axis and formed by two parallel rings, one solid and other with a mouth of width B that covers an ψ angle over the ring, its moment of inertia $I_{y(KR)}$ is two times the moment of inertia of a solid ring (Equation E.10) minus the one of the mouth (Equation

---

[47] Spinning at an angular speed ω, the angular momentum (L_s) of a body is defined as $L_s = \omega I$ where I is its moment of inertia about a spinning axis passing through its body and center of mass. For conservation of momentum ($L_s = constant$), when one of these (ω and I) variables increases the other decreases.

E.13) that should be subtracted because we are taking mass of the split key ring instead of adding. For details about this type of subtraction see Section *Balancing and the Add-cut theorem*. Hence

$$I_{y(KR)} = \rho A R^3 [2\pi - \tfrac{1}{2}(\psi + \cos 2\theta \sin \psi)] \qquad \text{(E.14)}$$

Where A is the transversal area of one ring, ρ the rings density, and θ is the position of the mouth or angle measured from the horizontal towards the spinning vertical y axis. When none of the two rings have a mouth, the angle ψ is zero and the moment of inertia given by Equation E.14 become twice (two solid rings) the value given by Equation E.10 for one solid ring.

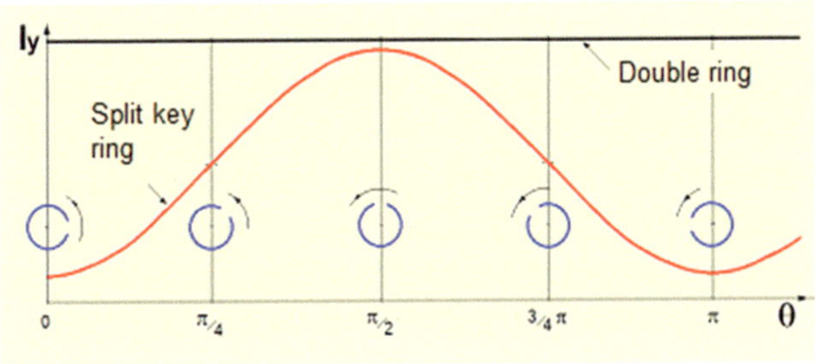

Figure E10. Variation of moment of inertia of a split key ring

Figure E10 shows the graph of Equation E.14 that associates how the moment of inertia of the split key ring varies with the mouth location on regard to angle θ. According to the graph, the moment of inertia of the split key ring spinning around a vertical axis will reach its minimum values when the mouth reaches the horizontal position at the farthest distance from the spinning axis (θ is 0 or 180 degrees). On the other hand, the moment of inertia becomes maximum when the mouth is at the spinning axis. That happens when the mouth is either at the top (θ is 90 degrees) or at the bottom section (θ is 270 degrees) of the ring. That seems not to agree with Equation E.7 which implies that the farther the mass is from the spinning axis the greater the moment of inertia.

That is true for an actual mass, but on the split key ring, the mouth does not have mass. It just represents the section of material omitted on one of two rings composing the split key ring. The closer the mouth from the spinning axis, the stronger its contribution to moment of inertia of the split ring. On the graph, the gap on values of moment of inertia between the peak of the split key ring and the hypothetical double ring is proportional to the lack of material of the mouth.

Approximate calculations about the Earth

Imagine you are traveling by car and are in an open region far from any town or city. You stop at a gas station for rest and to buy some groceries. When you fill up gasoline to your car you increase its internal energy. Once you hit the road again and reach the cruising speed part of such internal energy is being converted into kinetic energy or velocity. At this condition the car is consuming fuel at a steady rate to overcome the road friction and the air resistance to keep its velocity or kinetic energy. If at one moment you step on it (accelerate), this additional release of internal energy is transformed into more velocity and/or kinetic energy. When you do not inject gas or decrease its consumption, the car keeps its velocity due to its momentum for a short while. If the no-injection is sustained, the car starts reducing its velocity. To inform other drivers of these variations of energy, this future car is provided with back and front blue lights that are on when you keep the car traveling at constant speed, are off when you do not inject gas to the engine, and they increase their intensity when you step on the gas pedal. The conventional red light usually tells when the momentum or velocity is decreased abruptly.

Similarly to the car, in the Earth any change of its internal energy should be converted into another type of energy. Let's suppose that the Earth undergoes a change of its internal energy that is transformed into a change of its kinetic energy. Will this change of kinetic energy affect its orbital velocity, its rotational velocity, or both? If both, in what magnitude? Is one greater than the other? Let's suppose that this variation of energy is split into two equal parts, and one part becomes linear kinetic energy (K_L)[48] and the other rotational kinetic energy (K_R)[49]. If the equation of these energies are equalized, it is gotten that the linear velocity of the Earth is equal to its rotational velocity times its radius times a positive factor less than one, or in math terms $V = 0.63\ R\ \omega$. In other words, the linear velocity is much greater than the rotational velocity.

A similar relationship also can be obtained as following. Imagine our foregoing car traveling over a straight and flat road at a speed of *V* km/h. All persons and things in the car, and all its parts and components like the axle of its wheels are also moving at this velocity. However, in a non-slipping wheel its bottom point in contact with the surface of the *static* road also has the same zero velocity as the ground (*G*). The car or axle wheel *V* velocity can be represented by a horizontal vector as shown in Figure E11a, where the point *0* stands for the axle. Imagine that a line is drawn from the ground (*G*) and passing through the pointing end of this velocity vector, and it is extended to intersect a horizontal line drawn from the upper point (*U*) of the wheel. A vector from point *U* to the intersection point, represents the velocity of the upper (*U*) point of the wheel. This construction to get the velocity at point U is backward but equivalent to find an instantaneous center of mass (See explanation related to Figure A5a about this center). Notice that two triangles[50] were constructed, and they are of the same proportions. The horizontal side of the small triangle is vector *V* and its vertical side is the wheel radius *r*. In the big triangle its vertical side is twice *r* (or twice the vertical side of the small triangle), so its horizontal side shall also be twice or *2V*. This is

---

[48] $K_L = ½\ m\ V^2$, where *V* is the Earth linear velocity and *m* its mass.

[49] $K_R = ½\ I\ \omega^2$, where ω is the rotational velocity and *I* the moment of inertia that is equal to $2/5\ m\ R^2$ for a sphere of radius *R* and mass *m*.

[50] These triangles are similar because they are right triangles and they share the (bottom) angle at point *G*.

equivalent to say that the tire has two velocities, the car linear velocity and the turning one that makes the car travel. That is how one outside the car observes it. For instance when he/she is located on point *G* that is an instantaneous center.

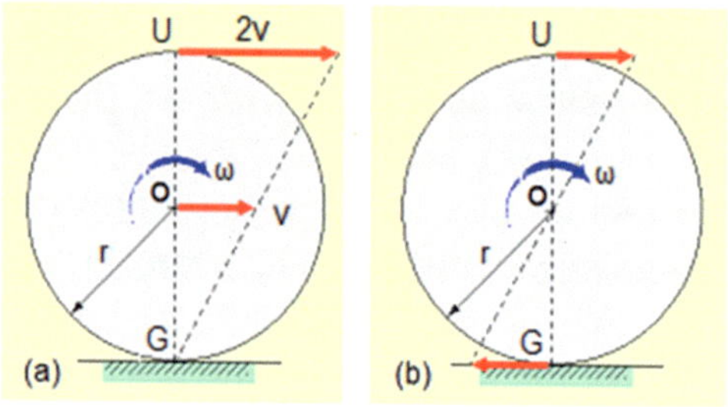

Figure E11. Car tire viewing from: (a) ground, (b) axle

From inside the car and when viewing to the outside through a small enough lateral window one does not appreciate the car traveling speed, instead one senses that the car is static and the external fixed objects seem to be the ones moving. To try this experience, place against the car-window glass the palm of one hand with the fingers contacting each other and the thumb separated of them. Do the same with the other hand which fingers and thumb shall also cover the first hand. Do it in such way to create a very small opening with the index fingers and the thumbs. Close an eye and place your face and the open eye against this hands-opening. You can generate the opening by other method. However, the important and key things are that your only view should be to the outside, and you should not see the car at all.

Through this small window, all external and fixed objects (like a tree or the point *G* on the ground) closed to the car seem to being moving at the same velocity as the car but on the opposite direction. If this viewing point or small window were on the wheel axle or point *0*, the observer would see not only the wheel upper point *U* moving at the car velocity on its same direction, but also the bottom point *G* (external object) but moving on the opposite direction. To visualize this result, imagine that the wheel is in the air for just a fraction of a second after jumping by hitting a speed bump. See Figure E11b. In this figure, it is equivalent that on the three points (*U*, *O*, and *G*) the velocity (*V*) be subtracted to the ones shown in Figure E11a. From the sketch of Figure E11b, one can infer that when the angular velocity (ω) of a rolling wheel is known, the linear velocity (*V*) of its geometric center is given by the converting relationship ($V = r\ \omega$), or vice versa.

This approach assumes that if a suspended wheel of radius *r* is spinning at ω rotational speed, once the wheel touches a fixed flat surface (ground), it will start traveling at the lineal velocity *V* calculated by the converting relationship ($V = r.\ \omega$). At low rotational speed that might be possible, but at high speed the wheel probably will start, for lack of traction, slipping over the same spot for a while. This happens with the wheels of cars competing on a drag race[51]. At the moment of the start when the engine is accelerated to its maximum speed, its traction wheels also reach their maximum spinning but on the same spot before the car starts going ahead.

The above relationship ($V = r\ \omega$) also applies to the round Earth, assuming it is spinning about an axis passing through its center of mass. In this case, if suddenly the Earth stops spinning, all that rotational kinetic energy (or angular momentum) is transformed into linear kinetic energy (or linear momentum) as if it were rolling over a flat surface. That assumption not only omits changes of internal energy, deformation, heat, etc. in the Earth, but also leaves out the fact that sudden changes of energy could be explosive and destructive.

Now let's find the ratio between the linear and rotational speed of the Earth using momentum. In general, the momentum of a body is defined as the product of its mass and its (angular or linear) distinctive velocity. That is, on a rigid-solid object of mass *m*, when it is moving at a linear velocity (*V*), its linear momentum

---

[51] A drag race is an acceleration race between cars usually over quarter-mile distance.

(p) is just $p = m\ V$. When the object is spinning around an axis passing through its center of mass at an angular speed ω, its angular momentum (L_S) is defined as $Ls = \omega I$[52] (Serway and Jewett 318), or $L_S = 2/5\ \omega\ m\ R^2$ when the moment the inertia (I) is evaluated for a solid sphere (see Equation E.13) like the Earth. Assuming that a variation of energy of the Earth is converted into momentum and it is split into two equal parts, and one part becomes linear momentum and the other angular momentum. After equalizing the equations of these momenta, it is obtained that the linear velocity is equal to two fifth, times square radius, times the angular velocity, or in math terms $V = 2/5\ R^2\ \omega$.

The three procedures used to estimate the order of magnitude between the linear and the angular velocities of the Earth when there is a change on its internal energy, gave different results. However, all of them coincide in that the linear velocity shall be greater than the angular one, if any. Therefore, when there is a change in the internal energy of the Earth, it is possible to assume that its lineal (orbital) velocity should be much greater than the angular velocity. This assumption is far from accurate, but it gives an educated estimate of the order of magnitude of this rotational velocity "transformed" into a linear velocity.

The actual values of the velocities of the Earth also seem to corroborate the above assumption. Using the relationship $V = r\ \omega$, and knowing that the Earth has a mean radius of about 6370 km, and revolves around its axis on one day ($\omega = 0.2618\,{}^{rad}/_{h}$), the equivalent linear velocity of its center is about 1,667 km/h (0.2618 x $6.37\text{x}10^3$) if it were on a pure rolling on a flat surface. We would get a value a bit lower or much higher if we used any of the other two approaches. On the other hand, the Earth-Sun distance (radius) is about $1.496x10^8$ km, so the length or perimeter of its orbit assuming it describes a perfect circle ($perimeter_{circ} = 2\pi r$), is about $9.39\text{x}10^8$ km. It means that the orbital speed of the Earth

---

[52] This equation is valid for a wheel and a sphere, like the earth, that are spinning about an axis passing by their center of mass.

is about 107302 km/h (9.39×10^8 km / 365 days). Thus, the actual orbital speed of the Earth is about 64.3 (107302/ 1667) times greater than its “transformed” and actual rotational speed. The result confirmed that the linear velocity of the Earth is greater than its rotational velocity.

Part F Review of mathematical topics

The roots of the present lie deep in the past and almost nothing in the past is irrelevant to the man who seeks to understand how the present came to be what it is.
Morris Kline

Great advances of the past may seem like a small thing nowadays. The boom of computer and electronic technologies surely will be considered insignificant soon, in maybe 50 years more! History teaches us that many topics which caused tremendous enthusiasm and captured the attention of the people ultimately faded into oblivion.

For example, the surface area of a rectangle was the starting point, to get the area of a triangle and trapezoids that were known by the Babylonians and Egyptians about 3000 B.C. The Egyptians knew the correct formula for the volume of the truncated pyramid. The Pythagorean Theorem caused a big sensation when it was presented for the first time (500 B.C.). Hundreds of years later, Democritus (460-370 B.C.) discovered the volume and surface area of a right circular cone. The volume of the cone inspired Archimedes (287-212 B.C.) to find the volume of a sphere, and the cousin of the cone, the frustum, helped him determine the surface area of the sphere. These pieces of knowledge, several centuries later, were the inspiration of the great Newton (1642-1727) to present to the world what is known as calculus. Mathematics not only is a language understood and shared by all ancient and current civilizations but also it is intrinsic and subtly embedded in other sciences and things of the world. Computers and electronic devices exist because of mathematics!

Deducing the equation of an ellipse

An ellipse can be constructed over a piece of flat wood using two nails as guides, a loop of twine whose length is greater than twice the distance between the two nails, and a pencil whose drawing point is kept inside the twine loop. That, during the construction, is kept tensed by the pencil point which is moved around the internal nails while drawing the ellipse on a paper attached to the wood or direct on it. Keeping the twine tense implies that whatever the position of the pencil point, it is always pulled from two directions aiming toward the two foci.

The origin of the coordinate's axes is usually located on the foci mid-point because in this position the two lines coinciding with the (X and Y) axes divided the ellipse into two symmetrical parts. Assume that the diameter of the nails is nil or just a point. The distance from the midpoint between the foci to a focus is identified as *c*, so that the distance between foci is *2c*. This location of the coordinate's axes allows for study two important pair of points or extreme ends of the curve.

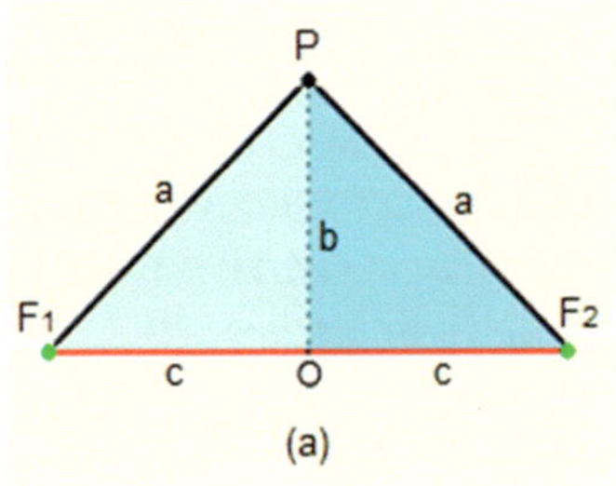

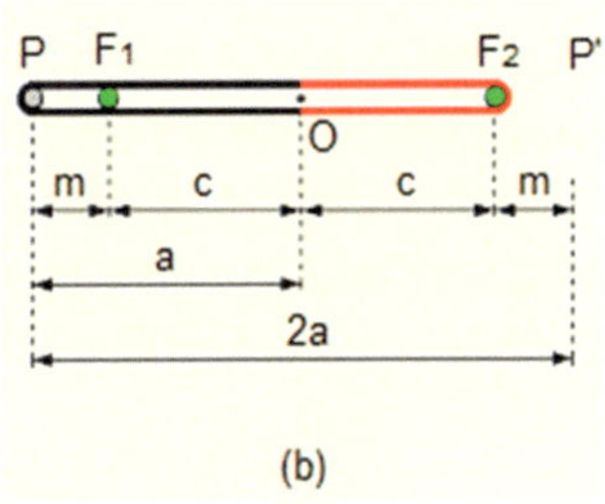

Figure F1. Extreme points of an ellipse

The first extreme pair is found when a point *P* is at the farthest distance from X axis, or when it is just in the front of the foci mid-point (O) reaching a distance *b* above (or under) the midpoint as shown in Figure F1a. In this position the two foci and the point form an isosceles triangle that is bisected into two rectangular triangles by the imaginary vertical (dashed) line joining the point and the foci midpoint. From one of these rectangular triangles, calling *a* the distance between the *P* point and one of the foci, and using Pythagorean Theorem, it is obtained that:

$$a^2 = b^2 + c^2 \quad \rightarrow \quad b^2 = a^2 - c^2$$

Let's call length L_o the tracing perimeter of the loop of twine used to draw the ellipse, and according to Figure F1a, this perimeter is essentially made up of two parts the active length or distance *2a*, plus the inactive section *2c*, thus

$$L_o = \underbrace{2a} + 2c$$

The second extreme end happens when a point of the ellipse is aligned with *X* axis and is simultaneously away from the two foci. This condition is met by two points, the nearest (*P* point) to focus F_1 and the nearest (*P'* point) to focus F_2. See top sketch of Figure F1b, where the diameter of foci and pencil are exaggerated for clarity. Let's call *m* the distance from one of these extreme points to its nearest focus (F_1 *or* F_2).

In this position the twine is divided into two halves having identical length, one above and the other below the nails, so the sum of them is the twine's overall length. According to the sketch, the length of each of these halves is equal to *m+2c*, so both or twice this distance is (2*m* + 4*c*) the twine length, or also:

$$L_o = \underbrace{2m + 2c} + 2c$$

As the twine length is constant, the first two terms of the right side of this relationship must be equal to the first term of the right side of the other L_o relationship gotten from Figure F1a, or:

$$2m + 2c = 2a \quad \text{or} \quad m + c = a$$

A graphical way to validate the above relationship: from Figure F1a, we know that the two up and slant (black color) sides of the triangle have a length of *2a*, and the horizontal bottom (red color) side has a length *2c*. When the bottom length (*2c*) is wrapped up from the midpoint o around the (right) F_2 focus, as it shown in Figure F1b, the remaining (black color) twine length *2a* has to be the one distributed into two identical parts, above and below the nail on F_1, from midpoint *o* to point *P*. Therefore, half of this value (*2a*) is *a*, or the distance (*m+c*) from foci midpoint *o* to point *P*.

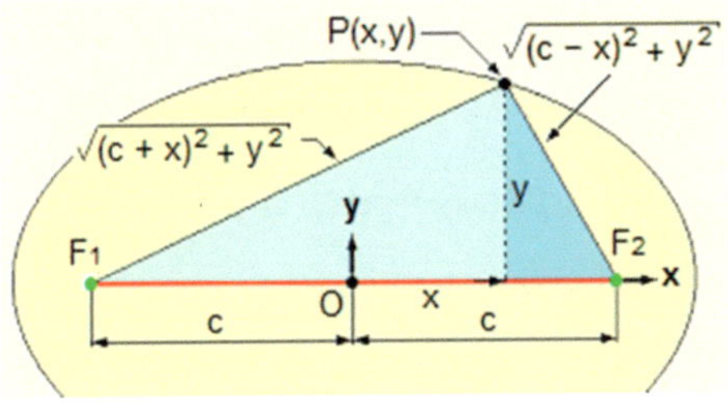

Figure F2. A generic point of an ellipse

As the twine length is constant (*Lo* = *2a* + *2c*) and a part of it (*2c*) is always between the foci, the rest of its length (*2a*) is allocated at any point of the ellipse. For that reason, this curve is defined as a set of points whose sum to two fixed points or foci (plural of focus) is constant. See Figure F2. Knowing that the two distances from an ellipse point to the two foci add to *2a*, it is obtained from Figure F2 that:

$$\sqrt{(c+x)^2+y^2}+\sqrt{(c-x)^2+y^2}=2a$$

After squaring (twice) and simplifying, we get

$$(a^2-c^2)x^2+a^2y^2 = a^2(a^2-c^2)$$

Using the relationship among *a, c*, and *b* ($b^2 = a^2 - c^2$), and dividing the above equation for a^2b^2, it becomes

$$\frac{x^2}{a^2}+\frac{y^2}{b^2}=1$$

The above equation is for an ellipse centered on the *XY* coordinates and which foci are aligned to the *X* axis. When the foci are aligned to the *Y* axis, the constants *a* and *b* interchange position on the equation.

The drawing and development of the ellipse equation highlights an important fact, specifically that to generate this curve a minimum of two foci are required. In other words, any point *P* of this curve is always withstanding two forces aimed towards the two foci.

The circle and the hyperbolic spiral curve

A circular segment is the remaining part of a circle limited by a chord that is the segment of a straight line intersecting the circle at two points. The arc of the circular segment is also limited by a central angle whose vertex is the circle's center, Figure F3.

An important theorem of geometry says that a diameter[53] perpendicular to a chord bisects the chord and its arc. Based on this theorem and by construction on the circle of radius R and center P, the line PM is part of the perpendicular diameter that bisects the chord 0Q and its central (4θ) angle. Similarly, the diameter through PT bisects perpendicularly the chord OM and its center (2θ) angle. The triangle P0M is isosceles by having two congruent sides (P0 = PM = R), and its central angle (2θ) at P vertex is bisected by the bisector ray PT, that is perpendicular to cord 0M too. This last bisection creates two new congruent and right triangles (P0T and PTM). On triangle PTM its angle at M vertex is 90-θ but this same angle on triangle 0NM is 90-β. So for this angle to be unique, angle β has to be equal to θ; or β = θ. That is also proof that the angle β or θ is a fourth of the initial central angle.

Based in such proof, it is possible to deduce (development not expose in this book) that the point M or the ray *r* (OM = r) describes a well-known curve, a Cochleoid whose equation is given by

[53] A diameter is a (line) chord through the center of a circle that divides it into two half or semicircles.

The circle and the hyperbolic spiral curve

$$r = \frac{L_O}{2}\frac{Sin\,\theta}{\theta}$$

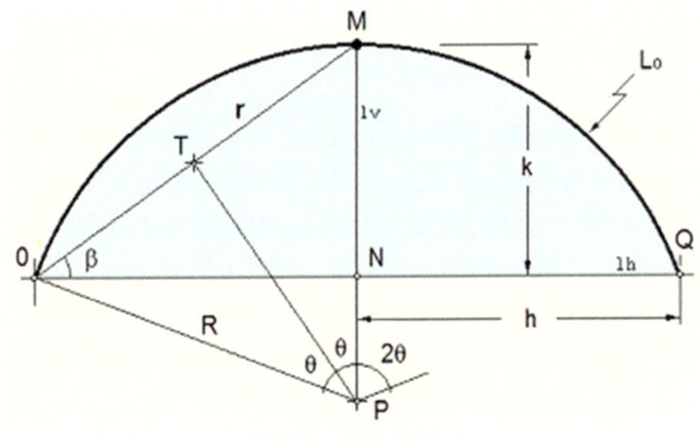

Figure F3. A circular segment

Let's assume that a hypothetical rectangular and flexible ruler of length L_o only bent into circular segments while its ends are forced to approach toward each other to a common point, meeting point, following a straight approaching line.

Let's find out the formulas to get the bent ruler midpoint height k and the distance h or half chord length needed to draw circular segments having the same length L_o, and touching a common point 0. Assuming that the point P was used to draw the circular segment shown in Figure F3, it is gotten that

$$h = 0N = Rsin2\theta,$$

$$h = Rsin2\theta \qquad \text{(F.1)}$$

also

$PO = PM = R,\ PN = Rcos2\theta$, and $k = PM - PN$, then

$$k = R(1 - cos2\theta) \qquad \text{(F.2)}$$

The arc (S) or the continuous part of a circle of radius R formed by a central angle (α, measured in radian) between two radii of such circle, is given by S = R.α. Based on this relationship, known as the circle arc-angle-radius, for an arc *S* equal to L_o and a central angle of 4θ, it is gotten

$$R = L_o/4\theta \qquad \text{(F.3)}$$

Figure F4 shows three circular segments, the one whose center is *P*, other (dashed line) with its center on *p'* point, and the other a circle that is formed when the ends of the ruler meet at the meeting point. We defined the last of these isoperimetric segments as *the guide circle* of radius r_o and center on point *S*. Regarding Figure F4, what is the curve generated by the trajectory followed by the centers, *P* points including *S* and *i*, of all circular segments of length L_o having their ends over the same straight approaching line and sharing one common (meeting) point *0*?

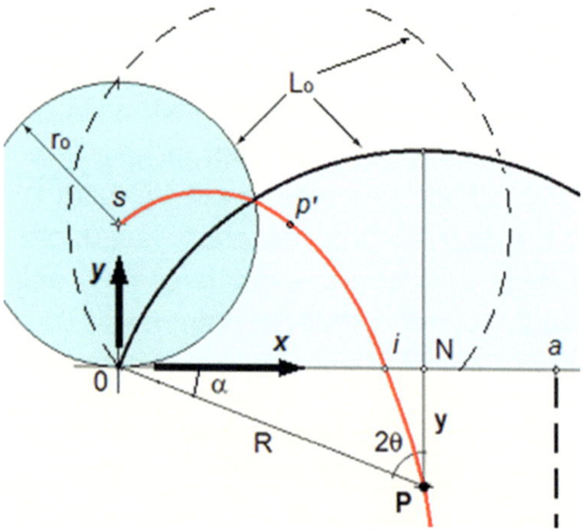

Figure F4. Trajectory of centers of isoperimetric circular segments

From triangle 0NP and adopting the convection that the α angle is positive when moves in clockwise direction from x axis

$$\alpha = \frac{\pi}{2} - 2\theta \;\rightarrow\; 2\theta = \frac{\pi}{2} - \alpha$$

Loading this relationship into Equation F.3 modified as

$$R = \frac{L_o}{4\theta} \rightarrow R = \frac{L_o/2}{2\theta}$$

And after arranging the terms, it is gotten that

$$R = L_o/(\pi - 2\alpha) \qquad \text{(F.4)}$$

That is the polar equation of the old and famous *hyperbolic spiral curve*[54], which is also known as *R.α = constant*. Also from triangle ONP we get

$$x = R\cos\alpha, \quad y = R\sin\alpha$$

And after substituting the value of R given by equation F.4, the above equation becomes

$$x = L_o\cos\alpha/(\pi - 2\alpha), \quad y = L_o\sin\alpha/(\pi - 2\alpha) \qquad \text{(F.5)}$$

The x and y relationships happen to be the parametric equation in Cartesian coordinates of the hyperbolic spiral.

The circumference (perimeter) of a circle is equal to Pi (π) two times its radius. For the *guide circle*, of radius r_o, whose circumference is the length (L_o) of the hypothetical ruler from which the circle would be formed, its perimeter is $L_o = 2\pi r_o$. See Figure F4. From this formula, the radius of the guide circle is:

$$r_o = {}^{L_o}/_{2\pi} \qquad \text{(F.6)}$$

For an angle α equal to minus 90 degrees ($-{}^{\pi}/_{2}$) input on either (F.4 or F.5) equation of the hyperbolic spiral the point "s" has a value equal to the radius (r_o) given by the above equation. Similarly, for an angle α equal to zero degrees is found the

[54] This curve was studied by Pierre Varignon in 1706.

intersection of the curve with the horizontal axis (point "*i*") whose *R* value ($R = L_o/\pi$) is two times the radius of the guide circle, or $2r_o$. A circle having a radius $2r_o$, twice than r_o, will have a circumference twice than the one of the blue circle of radius r_o, so half of this big circle (a semicircle) will have the same circumference (perimeter) as the blue circle. Thus, point "i" is the midpoint of the chord of a circular segment that is a semicircle of radius $2r_o$.

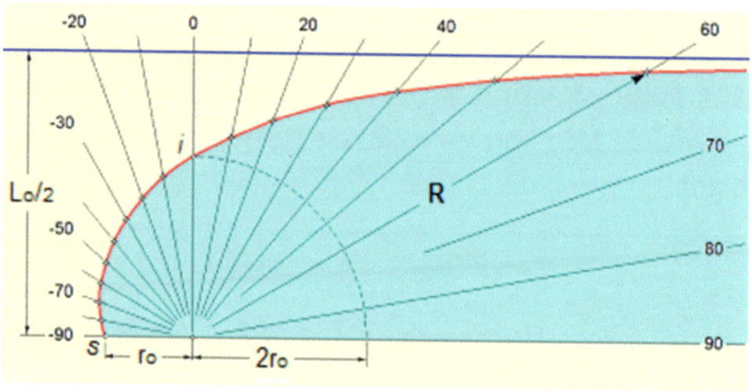

Figure F5. A hyperbolic Spiral curve

Once the drawing was completed, the coordinate axes were rotated to get the horizontal curve shows in Figure F5. In this curve notice that for angles greater than 70 degrees the radii R are so big that it becomes impractical showing more curve points. For angles of 80 degrees, and up the radii are far away to the right of the book page, and the curve is almost flat and almost parallel to the blue line. That is a limit which cannot be reached by the curve. It represents what is known in calculus as an asymptotic line or asymptote. In Figure F4, this line passes through point *a* and it is got when the ruler is flat, it means this point is the midpoint of the straightedge and for that reason it or the blue line is a Lo/2 from 0 point. However, in mathematics, it is said that the curve will (or seem to) touch its asymptote on the infinity. This "infinity" effect is like driving on a prairie, or on a desert over a straight road whose edges (or curbs) we perceive to be very close (almost touching each other), at a far distance in front of us.

The rpm of a hanging bicycle wheel freely spinning around its axle, decreases following a hyperbolic spiral. See section *The offset hanging spinning wheel* for details.

Part G Bibliography

Bombing Hitler's Dams. By Ian Duncan. Dir. Ian Duncan. Prod. Tom Cook. PBS Distribution, 2012. DVD.

Cox, Brian. *Galileo's Famous Gravity Experiment*. 18 March 2017. BBC Two. Movie. 2017. <Htps://www.youtube.com/watch?v=Qyef-_QPSbk.>.

Espenak, Fred. *Eclipses and the Moon's Orbit*. 12 January 2012. NASA. <https://eclipse.gsfc.nasa.gov/SEhelp/moonorbit.html>.

Genius of Britain - The Scientists Who Changed the World, Disc 1. Dirs. Tim Usborne, Christopher Sykes and Robert Bee. Perf. Stephen Hawking and Robert Winston. Prods. Tim Usborne, Christopher Sykes and Robert Bee. Acorn Media U.S., Silver Spring, MD, 2010. DVD.

Hawking, Stephen. *A Brief History of Time*. New York: Bantam Books Trade Paperbacks, 1998.

Kern, Donald Q. *Process Heat Transfer*. New York, NY: McGraw-Hill Book Company, 1965.

Kline, Morris. *Mathematical Thought From Ancient to Modern Times*. Vol. 1 & 2. New York: Oxford University Press, 1972.

Serway, Raymond A. and John W. Jr. Jewett. *Physics for Sciencists and Engineers*. Belmont, CA: Thomson - Brooks/Cole, 2008.

Spangler, Steve. *Naked Eggs and Flying Potatoes- Unforgettable Experiments That Make Science Fun*. Austin, Texas: Greenleaf Book Group Press, 2010.

Sykes and J.B., *The concise Oxford Dictionary of Current English*. Seventh Edition. New York: Oxford University Press, 1983 (reprinted).

Wikipedia. *Earth's Rotation*. 4 May 2017. <https://en.wikipedia.org/wiki/Earth's_Rotation>.

—. *Magnus Effect*. 10 March 2017. <Https://en.wikipedia.org/wiki/Magnus_Effect>.

—. *Orbit of The Moon*. 16 May 2018. <https://en.wikipedia.org/wiki/Orbit_of_the_Moon>.
—. *Solar Wind*. 4 May 2017. <https://en.wikipedia.org/wiki/Solar_Wind>.

Made in the USA
Las Vegas, NV
04 November 2026

57431528R00109